BURLEIGH DODDS SCIENCE: INSTANT INSIGHTS

NUMBER 90

# Sustainable tropical forest management

I0130675

burleigh dodds
SCIENCE PUBLISHING

Published by Burleigh Dodds Science Publishing Limited
82 High Street, Sawston, Cambridge CB22 3HJ, UK
www.bdspublishing.com

Burleigh Dodds Science Publishing, 1518 Walnut Street, Suite 900, Philadelphia, PA 19102-3406, USA

First published 2024 by Burleigh Dodds Science Publishing Limited
© Burleigh Dodds Science Publishing, 2024. All rights reserved.

British Library Cataloguing in Publication Data
A catalogue record for this book is available from the British Library

ISBN 978-1-80146-653-0 (Print)
ISBN 978-1-80146-654-7 (ePub)

DOI: 10.19103/9781801466547

Typeset by Deanta Global Publishing Services, Dublin, Ireland

# Contents

# Series list

# Chapter 1

## An overview of tropical forest formations

Alice Muchugi, World Agroforestry (ICRAF), Kenya; Sammy Muraguri, Kunming Institute of Botany, China; Hesti L. Tata, Forest Research & Development Centre, Indonesia; Jürgen Blaser, Bern University of Applied Sciences, Switzerland; and Patrick D. Hardcastle, Forestry Development Specialist, UK

## 1 Introduction

The type of forest encountered on any specific site on the planet is the result of the interaction between abiotic factors and the living organisms that together make up the ecosystem. This interaction is then further complicated by a range of influences that may be human; biological such as pollinators, pests and diseases; or mechanical such as wind, storms and wildfires. These influences operate over different scales, the largest of which is climate. However, climatic effects are themselves also modified by various effects and the scale of their variation. For the purposes of this chapter, we provide here a brief summary of the overarching complexity within which tropical forests may lie. Deliberately, the present chapter focusses on the biome 'tropical forest' and not the anthrome but being aware that the species *homo sapiens* has altered the biome tropical forests fundamentally over the past decades.

## 2 The tropics: defining the enabling conditions for tropical forests

The tropics are geographically defined as the region on the Earth's surface between 23°27′N (tropic of Cancer) and 23°27′S latitude (tropic of Capricorn). More accurately the tropics are defined in terms of climatic periodicity, including temperature, light (solar radiation) and hydrological periodicity. In the tropics,

http://dx.doi.org/10.19103/AS.2020.0074.02

the daily temperature fluctuations are higher than annual fluctuations. According to Troll (1961, cited in Lamprecht, 1990) the limits of the tropics are formed by the line of equilibrium between the daily and annual temperature variations. With respect to light periodicity, there is relatively little fluctuation in the duration of day and night. At the equator, day and night last 12 hours each throughout the year, and at the two tropics the longest day lasts 13.5 hours and the shortest 10.5 hours. In terms of hydrological periodicity there is no unique characteristic distinguishable for the tropics. As the sun lies directly overhead with high levels of solar energy, the tropics experience increased water evaporation as compared to temperate regions. Near the equator rising warm air is subjected to adiabatic cooling and daily rainfall occurs (Thomas and Baltzer, 2002). Varying climatic systems make the Earth's equatorial zone both warm and wet, thereby supporting the characteristic diverse flora and fauna of the tropical rainforest.

The relevant environmental site factors in the tropics relate to (i) temperature, (ii) precipitation, (iii) sunlight, (iv) atmospheric and soil chemistry and (v) mechanical factors.

## 2.1 Temperature

The low seasonal fluctuations are a characteristic of the tropics as a whole, including lowland (hot) tropics, mountainous areas (cold tropics) and also, although to a lesser extent, the dry tropics. What creates the particular condition for tree/plant growth and diversity is the constancy of the mean annual temperature over the year, independent of the temperature. In contrast, daily fluctuation of temperature can be important. For example, in Bogor, Indonesia, the difference between the warmest month (25.9°C) and the coldest month (24.9°C) is 1°C, while typical daily temperature fluctuation is 10°C from 23°C at 0600 hours to 33°C at 1400 hours. At higher elevations, the daily fluctuation increases further.

The high energy flux means the relationship between insolation and temperature is different in tropical and temperate regions. Locations in the temperate zone may have a similar mean temperature to a location in the tropics, but the annual temperature variation is much less in the tropics and the solar energy received is much greater. Consequently, the tropics experience increased water evaporation compared with temperate regions. The climate cycle is presented in more detail in Chapter 5, but, in essence, rising warm air is subjected to adiabatic cooling leading to rainfall.

## 2.2 Precipitation

The patterns of tropical precipitation are derived from the circulation of the air masses regulated by the position of the sun between the tropics of Cancer and Capricorn. At the equator, the sun is in a zenithal position twice a year (March and September). This is the time when the extreme hot and moist air rises in the atmosphere and cools off, creating convective clouds and zenithal rains. The clouds then travel southbound or northbound at high altitudes and descend at 30–40° latitude in the subtropical anticyclone and return to the equatorial cyclonic zone as the so-called trade winds. The intertropical convergence zone, which follows the sun being directly overhead, results in the trade wind patterns, and the moisture-laden air rising over land masses is subject to adiabatic cooling and consequent rainfall. Depending on latitude and topography the tropical belt experiences a perhumid climate essentially around the equator where rainfall exceeds evapotranspiration throughout the year, with some additional influence from soil depth and water storage capacity as well as topography on this.

The outlined process is fundamentally altered by the monsoon winds, in particular, on the southern and eastern borders of tropical Asia. During the northern summer, in latitudes between about 10°N and 23°N, this region is dominated by rain-bringing winds blowing into the equatorial cyclone that has shifted during the northern summer as far as to the northern part of India. The northeast trade winds blow only in winter when the pressure distribution is inverted.

From the equatorial belt the climate becomes increasingly seasonal and this seasonality is also affected by topography. Important alterations in the hydrological regime also occur locally and regionally due to a whole array of factors. Moisture-laden winds crossing high relief hold less moisture afterwards, and leeward areas will be progressively drier. The intensity of the resulting dry season is progressively amplified as the two peaks of the bimodal rainfall pattern of the equatorial zone merge into a single peak beyond around 10° latitude. The effect of altitude can overlay this simple model as the temperature drops with increasing altitude, and moisture is deposited by direct condensation as well as rainfall. The presence of cold ocean streams influences coastal vegetation (Peruvian Pacific coast). Another well-known regional effect has been well described in the case of Brazilian Amazon, where it was described that nearly half of the precipitation of the Amazonia originates from the evaporation of the forests in this area. The other half is from the Atlantic Ocean (Salati et al., 1978).

## 2.3 Light

As there is a relatively limited length of daylight throughout the tropics, tropical plants are adapted as short to medium day plants. This is also the main reason that temperate or boreal species cannot flower or fructify properly in tropical climates. Although there is intensive insolation due to the high position of the sun throughout the year, clouds and high humidity restrain the amount of light reaching the forest canopy. In the forests only the trees reaching the canopy layer and the emergent trees receive the full sunlight available. Light intensity can fall to 1% in the understorey (Lamprecht, 1990). Trees in that layer have a shade habit, including broad, loose crowns, horizontal branches and soft and large leaves.

Particularly in the humid tropical forest types, light is the main ecological factor. Most of the tree and plant species in the forest canopy are adapted to receive only limited sunlight. Regeneration of forests, however, needs horizontal opening of the canopy layer, and regeneration happens in the gaps of the forest canopy, large or small scale. When mature trees die and fall, the gap created allows increased light at the ground level and stimulates the growth of suppressed seedlings and saplings. If the gap is large, then it will be colonised by more light-demanding species and the succession process will retreat to an earlier stage. Where disturbance levels are low, few gaps will be created other than by occasional falling trees. The stable forests growing in such conditions are predominantly linked to hotspots of forest biodiversity (Whitmore, 1998).

## 2.4 Atmospheric and soil chemistry

While the chemical conditions of the air (atmospheric conditions) are relatively uniform worldwide, the properties of tropical soils differ essentially from the soil characteristics in temperate or boreal zones.

Tropical forest soils are generally old, not disturbed by major climatic changes in the past, with year-round high temperature and precipitation that has led to intensive weathering and leaching of nutriments, particularly in the humid tropics. Moist tropical soils have virtually no regenerative power. They lack essential minerals for plant growth and are generally composed of iron and aluminium oxides. The typical clay minerals of such ferralitic soils are kaolinites with low cation exchange capacities. In a largely undisturbed humid tropical forest, the exchange capacity is assumed by the humic substances, a fine organic compound and components of the humus layer. The humic substances are part of the thin layer, only a few centimetre-thick layer of humus which is produced by the litter mineralised within a few months. This system reinforced by the presence of mycorrhiza is adequate to maintain soil fertility in largely undisturbed or sustainably managed natural tropical forests.

Due to the extremely low fertility of tropical soils overall, tropical climax forests apply a series of strategies to protect themselves from nutrient loss. One of the core strategies is their diversity and the multi-storey structure of the forest stands. The specific forest structure intercepts nutrients from the air and rain. Also, the root system works as a filter through a web of fine roots. Mycorrhizae assure the complete uptake of the nutrients released in the process of humus mineralisation. As Lamprecht (1990) pointed out, in the humid tropics, 'the soils are maintained by the forests and not the forests by the soils'.

At the local and regional level, throughout the tropics, there are a number of areas with predominance of 'young' tropical soils, with high levels of nutriments, including volcanic andosols and alluvial soils in inundation areas of tropical Whitewater rivers, which bring a constant supply of minerals from young mountain ranges (such as the Andean mountain chain in the upper Eastern part of the Amazon or in the Mekong river basin on SE Asia). Tropical mountain areas also have more favourable soils with respect to nutrient contents. In the drier tropics, the soils are often nutrient rich (depending on the bedrock), but the limiting factor for both dry forests and agriculture is the availability of water.

Overall, about two-thirds (73%) of the soils in the tropical area are nutrient poor, including ferralitic and heavily weathered and leached soils, shallow as well as sandy soils (podzols), and about one-third of the soils have sufficient uptake of minerals, including fertile tropical black soils, alluvial soils and soils in semi-arid peripheral tropics (Weischet, 1980, modified).

## 2.5 Mechanical factors

Winds and storms (cyclones, typhoons and hurricanes), volcanic eruptions, lightning, forest fire, inundation and earthquakes influence the development of tropical forests. While some of these factors are unforeseen and considered as an 'unexpected' occurrence or 'surprise', today, increased vulnerability has been observed due to the effect of climate change, particularly the frequency and intensity of storms and forest fires.

Tropical storms can destroy intact natural forests, and there have been many cases over the past centuries that forests are destroyed before reaching climax, for example, 'hurricane' forests in the Caribbean and Central America or storm forests in Malaysia. Mangroves that naturally build a wall against flooding after storms or earthquakes/Tsunami have been increasingly threatened and partly destroyed over the past decades (e.g. Myanmar, Indonesia).

Lightning strikes are frequent in equatorial zones throughout the tropics, and they play a role in creating small gaps in tropical moist forests, generally favouring the regeneration of forests. Lightning strikes are also common in the subtropical belt.

Hail can be a problem in that it damages foliage and may facilitate disease damage by allowing fungi to colonise open wounds. Hail is infrequent at low elevations in the tropics, but at higher elevations, cooling in updraughts within cumulonimbus clouds is sufficiently severe for hail to reach the ground. The map in Cecil and Blankenship (2012) clearly shows the link between hail and attitude, with only high altitudes affected in tropical regions. Most literature on hail damage relates to plantations in the subtropics, such as Wingfield and Swart (1994).

Forest and landscape fires are of major significance as an ecological factor in all tropical forests, dry, semi-arid, semi-humid and today even rainforests, as, due to changing precipitation patterns and prolonged dry periods, the risk of vegetation fire has increased over the past 10-20 years. Some forest areas are considered as 'fire climaxes' (e.g. the lowland conifer forests in Central America and the extensive woodland savannas of East and Southern Africa) that have developed over many centuries due to repeated fires. In many areas throughout the tropics, the species compositions of moist and dry deciduous forests have altered due to frequent fires. While natural phenomena occur, most of the fires, today and in the past, are the result of direct human interventions, due to either uncontrolled agricultural burning, hunting carelessly or wilful infractions.

Results from the Ndola (Zambia) fire management plots started in the 1930s concluded that, while fire exclusion in *Miombo* woodland was infeasible over large areas, controlled early burning reduced the frequency and severity of fires. Until the 1980s, this practice was widespread in savanna woodland in Africa, both north and south of the equator. It subsequently fell into disuse as forest departments were increasingly inadequately funded, but its value is beginning to be recognised again (e.g. Hollingsworth et al., 2015).

# 3 Classifying tropical forest formations

## 3.1 Overview of the major tropical forest formations

The vast climatic variation and the specific conditions in the tropics result in a large variety of forest types, differing in their diversity, species composition and stand structure. There is no consistent, exact and universal definition of 'tropical forest', as changes are usually continuous and without sharp boundaries, depending on physical (climatic, edaphic, orographic) and biological conditions. The lack of a uniform classification of tropical forest formations is unfortunate because diverging definitions of forest types complicate comparisons between countries and continents (Corlett, 2016).

Tropical forests reach from moist deciduous forests (generally called tropical rain forests) to moist and dry deciduous forests, to subtropical, savanna and woodland, as well as into mountain ecosystems and specific edaphic forest ecosystems (e.g. peat swamp forests, heath forests, mangroves and tropical conifer forests). Table 1 gives an overview of the forest formations in the tropics, based mainly on climatic and particular soil characteristics.

Occasionally there may be an apparent mismatch between species composition and climate. One good example is the Nkhata Bay lakeshore in Malawi. The forest structure here most closely approximates tropical moist

**Table 1** Simplified classification of tropical forest formations

| Forest formations | Physiognomy | Climatic/site features | Main distribution |
|---|---|---|---|
| **Climax forest formations** | | | |
| Moist (humid) evergreen forests (tropical rain forests) | Evergreen, three- to multilayer forest canopy, rich in tree species | All months humid (>100 mm), very short periods without rain | Distributed from 10°N to 10°S of equator |
| • Lowland | Multi-storied canopy layers, trees with buttresses, cauliflory | Hot; $T_a$ 22–28°C $P_a$ > 1600–2000 mm | Amazon basin, Congo basin, SE Asia |
| • Montane | Three-storied canopy, rich in tree ferns and epiphytes (cloud forests) | Temperate/cold; $T_a$ 12–22°C $P_a$ > 1400–1600 mm, capture of humidity in mists/clouds | Andean countries, Eastern Congo, East African highlands, Upper Mekong, Borneo, PNG |
| Moist deciduous forests (semi-humid forests, semi-evergreen forests) | Part of tree species periodically deciduous, rich in tree species, 2-3 canopy layers | Alternating wet and dry season with dry season < 5 months | In continuum of evergreen forests generally in latitudes beyond 10-23°N and S |
| • Lowland | In rainy season forest appearance close to moist evergreen forest | Hot; $T_a$ 22–28°C $P_a$ < 1600–1800 mm | Central America, West Africa, Myanmar, India, South China, Sri Lanka |
| • Montane | Most upper storey trees without leaves in dry season, few epiphytes | Temperate/cold; $T_a$ 12–22°C $P_a$ < 1200–1400 mm, dry period cold | Mountain areas of Central America, SW Africa, upper Congo |
| Dry deciduous forests (mainly lowland and low montane) | All trees deciduous for part of the year except along watercourses and drainage lines. Single-layer canopy predominates | Alternating dry and wet season with dry season > 5–8 months Warm to hot; $T_a$ 17–25°C $P_a$ 500–1600 mm | Grading from moist deciduous forests as rainfall decreases generally in latitudes between 10-23°N and S: parts of Tropical America, East Africa, West Africa, India, Oceania |
| Dry forest savanna (woodlands in arid and semi-arid regions) | Periodically bare, trees widely spaced, fewer tree species, frequent occurrence of thorny species, xeromorphous structure | Long and harsh dry season, 8 to 10+ months Hot; $T_a$ 20–28°C $P_a$ < 500 mm | Sahel belt transition to desert, Southern Africa, dry part of India, parts of Australia |

| Forest formations | Physiognomy | Climatic/site features | Main distribution |
|---|---|---|---|
| **Special forest formations** | | | |
| Freshwater swamp and inundation, littoral forests, gallery (riparian) forests | Open to close forests, depending on the specific conditions | Rainforest climax on specific soils (swamp, peat); gallery forests in semi-humid to arid zones | Amazon, Guayanas, Orinoco, SE Asia mainly |
| Heath forests | Open forests, dominated by few tree species | Tropical rain forest climax on particular poor soils (podzols) | Amazon, SE Asia |
| Mangroves | Dominant in all tropical and subtropical coasts and river deltas and estuaries | Forests on specific soil conditions, in tropical humid, semi-humid and dry climates | Pantropical, on specific sites. Heavily threatened by development and climate change |
| Tropical conifer forests | Either dominant as conifer forests (Meso America) or scattered in climax hardwood forests | Semi-humid to perhumid climate, tropical mountain climate | Meso America and Caribbean, SE Asia; montane regions in Andes, East Africa and SE Asia |

forest, but the species composition is predominantly miombo savanna species. Chapman and White (1970) characterised it as 'evergreen dry forest'.

## 3.2 Moist evergreen forests (tropical rainforests)

Tropical rain forests are generally found mainly in lowlands (< 600 masl) in the equatorial belt in Central and South America, Central and West Africa, Indo-Malaya and beyond 10°S latitude in Australia (Zakaria et al., 2016). These forests receive a minimum of 1600-2000 mm of precipitation annually, and there is generally no dry period with monthly precipitation less than 80-100 mm. Out of the 10% of tropical forests' cover on the earth, the tropical rain forests occupy less than 7% but host half of the plant species on the planet (Tarakeswara et al., 2018).

Tropical rainforests (TRF) are structured in multilayers in a continuum, including sparse emergent trees reaching beyond the closed upper canopy, in an understorey with trees of different heights, and a relatively open forest floor. There are usually 300-700 trees >10 cm of diameter (at 1.3 m height) in 1 ha, with a canopy 20-35 m tall and emergent trees reaching 50 m of height and more and 50-300 species with more than 10 cm diameter in 1 ha (Malhi et al., 2002; Ter Steege et al., 2003). Other characteristics include many trees with relatively large leaves and often with large buttresses, alongside abundant palms, climbing plants, epiphytes and hemi-epiphytes.

These forests are evergreen, rich in vascular plant and tree species and in animal life. Particular structural characteristics that can be considered include canopy height and the presence or absence of the climbers and epiphytes. Other physiognomic features include tree buttressing, shape and layering of the crown, the position of the flower/fruit and the structure of the leaves. The highly complex structure of TRF has arisen from a very long process of progressive change that includes ecological succession in small gaps of the canopy, in which each stage creates altered conditions, particularly sunlight that results in changes to the species composition. In such gaps, in an initial stage, colonisation by short-lived, light-demanding species is observed and is thereafter progressively taken over by shade-bearing species until the forest becomes a stable climax community. Where disturbance levels are increased by natural means (forest fire, storms) or by human intervention (e.g. shifting cultivation), forest composition and structure changes, with pioneer or early succession species dominating over long periods of time rather than late succession species.

The ultimate stable climax tree species in wet tropical forests commonly produce hard and heavy wood, often with inclusions of silica or calcium carbonate. Thus, many of the tree species of moist tropical forests are not commercially harvested. Nonetheless there is a considerable number of timber species from not only moist evergreen forests but also moist deciduous forests that provide many of the world's commercially harvested hardwoods such as the many species of the family of the Dipterocarpaceae, dominating in large parts of SE Asian rainforests (*Shorea* spp., *Dipterocarpus* spp., *Hopea* spp. among others), African mahogany (*Entandrophragma* spp., *Khaya* spp., *Afzelia* spp.), harvested mainly in the countries of the Congo Basin and American mahogany (*Swietenia macrophylla*) growing from Mexico down to the Amazon basin, besides a large number of other hardwood species. Tree species from these moist evergreen forests in tropical America and tropical Africa are predominantly intermediate succession species rather than climax species, and they are thus found only in few numbers in closed moist evergreen and moist deciduous forests. This is in contrast to Asian rain forests, where trees of the family of Dipterocarpaceae dominate the stands.

A specific type of rainforest is the montane rain forest that stretches between 600 and more than 3000 m of altitude. Humid tropical montane forests ('cloud forest or mist forests') are found in almost 60 continental countries and many tropical oceanic islands (Aldrich et al., 2000) such as the Andean forests (from Costa Rica, Panama, Colombia, Venezuela, Ecuador, Peru to Bolivia), mountain belts of East Africa and in Central Borneo. The mountain rainforests vegetation composition changes with altitude due to adiabatic cooling and other factors such as wind exposure and varying soil characteristics. Depending on elevation, mists have a major influence on these forests due to condensation and reduced evapotranspiration (Stadtmueller, 1987). In upper montane forests, beyond 2500 m of altitude, especially on sites with poor nutrition, the trees tend to be short in stature and can be as short as 1.5–1.8 m. This is the elfin forest where the canopy has small and thick leaves attached to gnarled branches (Roberts et al., 2016). However, in some mountain regions (Costa Rica, Panama, Colombia) with deeper and more fertile soils where tropical oak species (*Quercus* spp.) dominate, trees can grow up to 45 m and at an altitude above 2500 m. Montane rainforests are less rich in species and simpler in their structure than lowland rainforests.

### 3.3 Moist deciduous forests (semi-deciduous forests)

Moist deciduous forests tend to alternate with the lowland evergreen rain forests depending on the average annual period of drought (Prideaux, 2014). During the dry period, several tree species, at least in the upper storey of the canopy, shed their leaves in up to one-third of the upper canopy. This forest formation represents a broad intermedium spectrum of forest types between evergreen rain forests and deciduous dry forests and occurs in a continuum between perhumid and dry climates. In Africa they occur in West Africa and adjacent to the rainforest in the Congo Basin, mainly south and east, and in Asia on the Indian subcontinent and SE Asia; in Tropical America they extend south and north to the Amazon rainforest and at the Atlantic side of Central America, and at the northern and eastern cost of Australia. As such, moist deciduous forests cover a potential area of about 250 million ha throughout the tropics (Lamprecht, 1990).

Comparing the semi-evergreen tropical forests to the evergreen rain forests, the species diversity in the latter is higher. Moist deciduous forests generally have a lower number of trees per ha than rainforests, and the understorey of the forests is denser with a higher number of small trees and ground vegetation. Against rainforests, the flowering and fruiting habits of most species follow the pattern of the two seasons (humid and dry). Fruits generally ripen at the beginning of the rainy season. Many of the main tropical timber species also grow in semi-deciduous forest formations, particularly those that are light demanding in a certain period of their lifetime, such as *Swietenia macrophylla* and *Cedrela odorata* in Tropical America and *Terminalia superba* and *Triplochiton scleroxylon* in West Africa.

### 3.4 Tropical dry deciduous/monsoonal forests

Dry deciduous forests are the third main climax forest formation found in the tropics. They occur mainly in the outer part of the tropics (beyond 20°N and S and extend beyond the tropics in regions in which potential evapotranspiration is much higher than precipitation. At its limits, it gives way to dry forest and thorny savannas, succulent steppes and semi-deserts.

The open 'dry forests' are characterised by loosely stocked, mainly deciduous forests with one to two stories, relatively poor in species. Most canopy trees shed their leaves in the dry season, which allows light to penetrate to ground level where the savanna ecosystem of grass thrives (Thomas and Baltzer, 2002).

Tropical dry forests intergrade with closed forests in Central and South America, much of Africa, including western Madagascar, South Asia (India) and parts of Oceania. In Central America, dry forests are found in the rain shadow of the mountain ranges extending from Mexico to Costa Rica. In South America, dry forest extend in the south-eastern part of Ecuador and north Peru in the Pacific region and in Venezuela and north-east of Brazil at the Atlantic. In East Africa, the physiognomy of dry forests is often characterised by pinnate-leaved species, for example, *Acacia* savanna and *Brachystegia* in miombo woodlands.

In South Asia and Africa (Sahel and East Africa), in particular, these forests are heavily populated by societies that depend on them for products such as fuelwood as well as non-timber products, for example, medicinal plants, fruits and fungi. Thus, tropical dry forests have been subjected to extensive conversion for cultivation and livestock wherever they occur.

Surprisingly, dry forests may also contain important timber trees, which may be naturally well formed, such as teak (*Tectona grandis*) in Myanmar, sal (*Shorea robusta*) in the dry monsoon zone of the Indian subcontinent and *Pterocarpus angolensis* in the savanna zone of East and Southern Africa with *Pterocarpus erinaceus* in West Africa.

## 3.5 Tropical forests under special site conditions

Apart from the main climatic forest formations, there are a number of specific forest formations developed under specific conditions, for example, in areas where the soil–water regime is distinct and thus creating forest formations that are distinct from the climax. Freshwater swamp forests, peat swamp forests, inundation forests, littoral forests, heath forests and mangroves fall under this category.

### 3.5.1 Peat swamp forests, freshwater swamp and inundation forests, gallery forests

These types of azonal forest formations are characterised by the dominance of the water regime that determines the forest formation in a specific climatic area. There are mainly two formations that can be distinguished: peat swamp forests on nutrient-poor substrates and freshwater swamp and inundation forests with moderate to high nutrient availability. Peat is an accumulated and non-decomposed plant material (Comeau et al., 2014).

Peat swamp forests are found in areas or regions with restricted drainage along rivers and may be localised or very extensive. The island of Borneo has extensive areas of tropical peat. In tropical America, remains of peat swamp forests appear in the Orinoco Delta and the Guianas (so-called Igapó), while in Africa this type of forest is nearly absent.

Peat swamp forests vary widely from small open forests on deep peat to high-closed dipterocarp forests dominated by species like *Shorea albida* in Borneo, which potentially can reach 60 m of tree height. Such closed forests can be observed in Brunei-Darussalam, whereas in other parts of Borneo, this forest type has disappeared and be replaced often by oil palm plantations developed on drained soils.

Research on peat swamp hydrology conducted between 1998 and 2001 in Kalimantan on the drainage of peat for a mega rice project and oil palm found counter-intuitive results in that peat depth increased by moving away from the drainage line due to perched water tables. It also found that there were two distinct ages in the peat with a gap of several thousand years across the unconformity. It became apparent from the research that the management practices being applied were inappropriate. Drainage for rice ran counter to the fundamental hydrology of the system, while forest clearance impacted severely on the water table and prevented the establishment of crops such as oil palm. Furthermore, degrading the forest canopy through overcutting was shown to be instrumental in the widespread fire damage and consequent loss of biodiversity (Page and Rieley, 1998; Page et al., 1999).

It is salutary to note that this facilitated the calculation of the catastrophic carbon emission from the fires associated with the mega rice project, which was estimated as some 30% of the total annual global carbon emissions from fossil fuels (Page et al., 2002, 2004).[1]

Freshwater inundation forests are found in low-lying locations near rivers that are occasionally flooded by river water, in some areas up to six months. Inundation causes the forest to have high soil nutrient levels that promote rapid growth of trees, although the waterlogging causes forest disturbance. Such forests can be found all over the flood plains of large rivers throughout the tropical lowlands, particularly in the Amazon basin, and also in the Congo Basin and tropical Asia. Freshwater inundation forests support many livelihoods in tropical moist forests as they are low lying, reachable and generally apt for agriculture. In the Amazon, native communities have been using and modifying these forests in which they have lived for thousands of years. Several studies suggest that the presence of nutrient-rich 'black earth' in these low-lying forests indicates intense fertilisation and burning of these areas (German, 2004).

Gallery forests (riparian forests) are the last group of azonal forests that grow under specific soil conditions and water regime. They often consist of narrow strips of forest running along the fringes of watercourses and the banks of lakes, in semi-humid, semi-arid and arid climates. The most important site factor is the constant supply of groundwater. Gallery and riparian forests have important protective functions but are also in many regions of the world the only available source of timber.

### 3.5.2 Heath forests

Heath forests occur on nutrient-poor, highly acidic, free-draining podsols that develop in hot and humid lowland climates on acid silicates, sandstones, sandy and marine sediments. In contrast to swampy soils, the key site factor is not the water regime, but the poorness of nutriments in the soils. In SE Asia they are known as 'kerangas', in tropical America as 'caatinga' or 'campinas'. Virtually no heath forests are found in African rainforests. Heath forests have a uniform open canopy with little vertical stratification. In many cases, tropical conifer species are main species dominating heath forests (e.g. *Agathis borneensis*

---

1 Note in terms of above-ground carbon stock, the highest levels are in temperate rainforests (Keith et al., 2009).

or *Podocarpus* species). In the Eastern Amazon (Guyana), the valuable timber species wallaba (*Eperua falcata*) occurs as the dominant species of the heath forest type. Heath forests often contain insectivorous plants, as may be found also in some lowland tropical forests growing on impoverished soils.

### 3.5.3 Mangrove forests

Mangroves occur worldwide throughout the tropical belt and in certain coastal areas in the subtropics, generally between the latitudes 25°N and 25°S. There are no comprehensive data on the total extent of mangrove forests. Indicatively, in 2015 the extent is about 13 million ha although data vary according to the source (Hamilton and Casey, 2016). In 1997, Spalding et al. (1997) estimated the world's mangrove area as 18 million ha. Major drivers of mangrove loss are conversions to aquaculture, agriculture and urban land uses (Spalding et al., 2010). WRI (2015) indicates the loss of mangrove area between 2001 and 2012 to be 192 000 ha.

Mangroves are forests situated along the coastal areas over much of the tropical region and may extend significantly inland, where the topography is very flat as in the Sunderbans in Bangladesh and India. The water in mangrove forests is salty or brackish, and the flora is extremely impoverished. The plants in these forests have special adaptations, such as the ability to retain high salt concentrations in their tissues enabling the plants to absorb saline water (Alongi, 2014). The Rhizophoraceae family have stilt roots and pneumatophores extending above the high-water level to allow the transfer of oxygen to their rooting systems. Globally, mangrove forests are characterised by the occurrence of a limited number of species, mainly from the family of Rhizophoraceae (*Rhizophora* spp., *Bruguiera* spp., *Ceriops* spp.), Avicenniaceae (*Avicennia* spp.) and Sonneratiaceae (*Sonneratia* spp.).

Mangroves provide an irreplaceable habitat and nursing ground for numerous animal species and are thus crucial for the livelihoods of people who depend on fisheries for income. To a lesser extent, local communities also collect timber as well as fuelwood for house construction. The importance of mangrove forests has increased with climate change. They store high amounts of carbon (Hamilton and Friess, 2018) and have important protective functions in view of sea-level rise and the occurrence of extreme weather events. They are also an important safety net for people living along tropical shores.

### 3.6 Tropical conifer forests

Tropical forest formations are largely dominated by broadleaved trees and forests (Blaser et al., 2011). Out of more than 50 conifer genera worldwide, at least 20 occur in the tropics; some of them are exclusively in the tropics (e.g. Agathis). Conifers developed in the carboniferous period under tropical climatic conditions in the northern hemisphere and migrated in later stages via the Cordilleran and Indo-Pacific bridge into America and Asia. Due to the lack of such 'bridges', tropical Africa is poor on indigenous conifer species with natural occurrences of four genera: *Cupressus*[2], *Juniperus*, *Podocarpus* and *Widdringtonia*, and this last genus is found only in Southern Africa.

Conifers are generally distributed throughout the tropics but more frequently outside the equatorial zone (e.g. *Agathis* spp., *Araucaria* spp. and *Cunninghamia lanceolata*) and more frequent in mountain forest areas than in the lowlands (the majority of the tropical pine species and *Podocarpus* spp., for example). *Podocarpus* is the only pantropical genus. *Pinus* spp. occur naturally in Central America and also in SE Asia.

The overall area covered by forest formations dominated by conifers is about 30 million ha, with an estimated 20 million ha in Central America and the Caribbean and 8 million ha in SE Asia and about 1 million ha in Africa (mainly dry-montane *Juniperus* forests). More than two dozen pine species dominate large tracts of climax forests in montane Central America and littoral forests in the Caribbean. In SE Asia, conifers occur more commonly in mixed forests. From about one dozen *Pinus* species that are native in the tropical regions of Asia, most of them grow mixed into broadleaved forests or occur as a dominant tree in areas of poor soils or in montane regions.

Tropical pine species, besides tree species from the genera of *Eucalyptus* and *Acacia*, have played a major role in tropical forest plantation development throughout the tropics, and, in particular, in East and Southern Africa. Among the most important species used for industrial forest plantations are *Pinus caribaea*, *P. oocarpa* and *P. patula* from Mesoamerica and *P. kesiya* and *P. merkusii* from SE Asia.

# 4 Structure and diversity of tropical forest ecosystems

## 4.1 Structure

Tropical moist forests are biologically richer than any other plant community (Tarakeswara et al., 2018), and many hypotheses have been advanced to explain this high diversity (Wright, 2002). The diversity of plants in the tropical rain forest is driven by the interaction of various elements such as seed predators, pollinators, pathogens, seed banks, seedling survival, light conditions and ant–plant interactions (Kakishima et al., 2015). Species diversity

---

2 Only *Cupressus dupreziana* occurs within the tropics and is critically endangered - Lábusová, J., Konrádová, H. & Lipavská, H. The endangered Saharan cypress (*Cupressus dupreziana*): do not let it get into Charon's boat. *Planta* 251, 63 (2020). *Cupressus atlantica*, also critically endangered, occurs in the Atlas Mountains in Morocco and Algeria, together with critically endangered *Abies numidica*, plus *Cedrus atlantica* and *Juniperus oxycedrus*. *Cupressus sempervirens* has a small population in Libya.

varies across locations primarily because of the distinct biogeography, habitats and disturbance levels.

Tropical moist forests also have the most complex canopy structure of any forest ecosystem, which creates the greatest diversity of ecological niches to support biodiversity. The greatest tree species diversity appears to lie in South America with 307 species with a diameter of 10 cm or greater recorded on a 1-hectare plot in Ecuador and 283 in Peru, both in the upper Amazon region. In Africa, the highest number recorded was 138 on 0.64 ha in Korup, Cameroon, which is a Pleistocene refuge. In Asia, the highest number of tree species recorded is in Kalimantan, with 240 tree species on a plot of 1.6 ha (Whitmore, 1998).

The most diverse communities in terms of flora and fauna are found in tropical rainforests. Floral composition is dominated by broadleaved trees with large buttresses that are covered with climbers, epiphytes and hemi-epiphytes. The canopy is also multilayered with diverse flora and fauna. According to the study by Ifo et al. (2016) on tropical forests in the Democratic Republic of Congo, the main floral groups present in the region are Euphorbiaceae, Fabaceae-Mimosoideae, Rubiaceae and the Guttiferae. They further noted that the floral composition is determined by both geological substrate and climate diversity. Zakaria et al. (2016) examined tropical forest ecosystems' floral communities by utilising the four distinct layers of the above-ground structure found in wetter tropical forests. These four distinct layers are:

**Emergent vegetation layer:** The topmost level consists of the emergent vegetation layer that is usually comprised of evergreen, broadleaved trees. Their trunks usually have a girth of 4-4.8 m with a height of between 30 m and 76 m. Climatic conditions at this level vary depending on the temperature and wind speed above the canopy level.

**Canopy layer:** This is the second and main layer of the tropical rainforest canopy. It is thick and densely filled with tree crowns and species such as rattan palms, philodendron, orchids, ferns and lichens. The height is between 18 m and 27 m from the forest floor. The immense structure and tree species diversity attract a wide range of fauna species - mammals, amphibians, arthropods and reptiles.

**Understorey layer:** This consists of small trees, shrubs and ferns that grow about 4 m high. Fungi, mosses and algae coexist in the trees that also attract a rich diversity of arthropods including bees, ants, butterflies, beetles and stick insects, which serve as a food source for birds and reptiles living in the forest.

**Forest floor:** This is the lowest layer. It is usually dark since only some 2% of sunlight penetrates to the floor (Roberts et al., 2016). Due to the limited solar energy, few plant species are found, mainly fungi and mosses.

However, it is rich in fallen leaves, seed, fruits and branches that make up the organic matter layer on the soil surface. Strong shade-bearing species seedlings may also occur here. Animals such as gorillas, elephants, jaguars, leopards, pigs and mongooses are found on the forest floor, although some also climb into the canopy layer to feed and sleep.

Forest structure becomes less complex primarily in response to decreasing rainfall and water availability, which also increases the incidence of fire. Referring to Table 1, moving from tropical moist forest to dry forest savanna, the number of tree species reduces, the canopy structure is simplified – often to a single layer – and the individual trees become increasingly more widely spaced until, ultimately, each may be isolated from its neighbours, at least above ground.

According to Tovo et al. (2016), the different species in diverse tropical forests display similarities regarding the population distribution across each species. This species diversity serves as a reservoir of new opportunities and responses that are transferred from one generation to the next, allowing them to maintain their functions and survive in a changing environment. Molecular tools have shown that most tropical tree species retain high genetic diversity within their populations (Muchugi et al., 2008) that may result from their outcrossing breeding systems.

In ecosystems with high species diversity, it has been found that interspecific and intraspecific diversity plays a crucial role within the ecosystem. The study conducted by Hazard (2018) found that interspecies diversity promoted increased interaction between the trees and organisms such as fungi that perform key processes such as nitrogen cycling in roots. These processes enhance the growth of trees and their nutrient acquisition. However, there was little indication that intraspecific species diversity was important, which could imply that it had minimal effect on the functioning of the trees.

Functional relationships within an ecosystem are driven by the different traits within the plants that form a community (Hahn et al., 2017). Wetter tropical forests have a richly diverse tree flora, which promotes their ecological roles and enhances their growth. However, Umaña et al. (2016) argue that, while tropical forests have strong intraspecific negative density dependence, this is mainly influenced by seedling assemblages although these only represent a small proportion of the total functional volume.

Variables such as leaf size also influence the intra-interspecies diversity within tropical forests because the trees have to adapt and survive in the area by competing intensively for water, light and nutrients. Where tropical forests encounter alternating rainy and dry seasons, chemical variation among plants is evident at different stages of growth and development, which affects the functional traits of trees (Sedio et al., 2017). These various studies imply that a relative importance exists across the intra-interspecies diversity spectrum and the trees' performance in the tropical forest ecosystem is dependent on how the various species relate to one another.

Nguyen et al. (2018) focussed on analysing whether stochastic effects dilute species associations in highly diverse communities, thus weakening them and whether clearly defined patterns occur mainly in local neighbourhoods. They found that intra-species distribution trends are controlled by dispersal limitation and habitat variability. Mutualistic interaction also contributes to tree species diversity and is common in tropical rainforests (Kakishima et al., 2015). Most trees found in the tropical forests often produce nectar and pollen that attract insect and vertebrate pollinators. Many tree species also produce fruits that are consumed by animals that act as seed dispersers. Angiosperm trees in tropical forests generally have mutualistic relationships with their pollinators and seed dispersers.

Despite the widespread recognition of the high tree species diversity in wetter tropical forests, no conclusive explanation for this has yet been deduced. Howe (2014) provides several threads of argument that species diversity in tropical forests arises from the rare species that remain unknown. First, the rare tree species in tropical forests exist because of the edaphic adaptations, seasonal or demographic niches and displacement of competing species. Secondly, Howe applies the neutral theory that argues that the distribution of trees within the tropical forest occurs randomly. Lastly, he considers the relative roles of chance and determinism; for instance, many common tropical tree species in Panama have limited seed dispersal despite having many common dispersal agents.

A study of 1400 tree species in wetter tropical forests in Panama, Ecuador and Colombia found between 36% and 51% of tree species are strongly associated with soil-nutrient distribution (Howe, 2014). Understanding rarity is further complicated because some species may be common in certain places but sporadic elsewhere. Tovo et al. (2016) used the relative species abundance (RSA) tool to determine the commonness and rarity of species in an ecological community. They concluded that, when the ecological factors or environmental conditions change, some of the rare species are more capable of surviving than others. Therefore, the diversity of trees within tropical forests appears to depend on the capacity of different species to evolve as the environment changes and their ability to replace competitors that cannot keep pace with these changes.

## 4.2 Dynamics

The great complexity of ecological niches within moist tropical ecosystems leads to complex faunal communities attracted to the diverse territories available to meet their food, shelter and reproductive needs. The various elements within this complex assemblage may be symbiotic, such as with pollinators, or it may be simple co-existence. The biological assemblage of an ecosystem includes microflora and fauna that may aid nutrition, as with mycorrhizal fungi, and/or breakdown of dead material as part of nutrient recycling.

Understanding the complex relationships within an ecosystem is important for the conservation and management of these ecosystems as well as for the use of individual species for plantations. Troup (1932) included a long speculative discussion on why pine plantations had failed in East Africa but succeeded in South Africa. The answer proved to be that fungi with which the pines could form mycorrhizal relationships were present in South Africa but not in East Africa. Once this was known and remedied, the trees survived well.

Maintaining viable forest ecosystems requires an appreciation of requirements for regeneration. Pioneer species require full light for seed germination and seedling establishment and can colonise open ground. Other species display varying ability to germinate and establish in the shade, and this characteristic is the basis of succession. More complex ecosystems, such as tropical moist forest, have developed through succession towards a stable climax forest in which gap size is important in determining the species composition (Brown, 1993). A large gap will become colonised by pioneer species, while small gaps will perpetuate shade-bearing climax species with the gap switch size determining the outcome.

Fragmentation, particularly of complex forest ecosystems, may result in only some components being present. Thus, even though mature trees may remain in the fragments, their symbiotic partners may not be present and ultimately the fragment will degrade. The minimum population size is an important consideration. While it varies for individual species a commonly accepted proxy is the 50/500 rule. This means 50 individuals are the minimum to avoid extinction, while 500 individuals form a viable population. In order to be an effective population, there has to be connectivity amongst these groups of individuals. Pioneer species often rely on wind dispersal, whereas late succession species tend to have much less dispersal distance capability.

Regeneration systems vary between different forest types. Most broadleaved tree species have some capacity for vegetative propagation by coppicing or suckering from roots. While seed is the predominant method for moist forest species, dry zone species where fire is a major ecological influence tend to rely more heavily on vegetative methods. In the Miombo woodlands of East and Southern Africa and similar fire climax ecosystems, there is great capacity for vegetative reproduction and root fragments have long viability. Consequently, some 90% of regeneration is vegetative. A comprehensive guide is available at: https://www.ltsi.co.uk/wp-content/uploads/2015/02/LTS-Regeneration-Potential-Guidebook-FINAL-non-print.pdf.

There is a range of other fire responses that includes vegetative and seed regeneration and fire damage reduction mechanisms. Lignotubers are found in numerous *Eucalyptus* species (http://keyserver.lucidcentral.org:8080/euclid/data/02050e02-0108-490e-8900-0e0601070d00/media/Html/index.htm and Jacobs, 1979) as well as in species such as *Pterocarpus angolensis*. *Eucalyptus* species have various fire resistance mechanisms including very thick bark to protect from fire and smooth bark to limit fires spreading to the crown. They also have vast numbers of dormant buds within the bark that will be stimulated into growth after fires. Many are also prolific seed producers - often of small seed - and display an ash-bed response, which allows germinants to grow very rapidly on bare ground with vast quantities of nutrients in the post-fire ash (Pryor, 1976).

Successful, rapid establishment of *Eucalyptus* mimics this situation to capture the ash-bed response by providing a weed-free site and applying high levels of fertiliser.

Some pine species have a grass stage where leaves protect the growing bud from fire while the root system is developed. Species include the *Pinus montezumae* group, some provenances of *P. merkusii* from origins where fires are frequent and *P. palustris* from the southern United States, again where fires are frequent. While most conifers do not coppice, there are some that do so. These include *Sequoia sempervirens*, *Cunninghamia lanceolata* and tropical/subtropical pines including *P canariensis*, *P. oöcarpa* and *P. roxburghii*, all from fire-dominated ecosystems.

## 5  Conclusion

The remaining area of tropical forest formations not visibly affected by human activities and with their ecological processes not significantly disturbed is dwindling rapidly. Some of the forest formations described briefly in this chapter have literally disappeared completely or have been reduced to small patches of forests, such as the Brazilian Atlantic rainforests or Kerangas peat swamp forests in Borneo, once dominated by *Shorea albida*. The area of so-called original tropical forest landscapes has shrunk to less than 500 million ha (ITTO, 2020). Thus, the extent of intact forest formations that regenerate naturally with the native species at the site with a closed nutrient cycle and allowing natural processes of succession is reduced by nearly two-thirds of its initial extent (ITTO, 2020). In these remaining areas of intact forest formations, depending on the scale and frequency of natural influences causing a disturbance, the composition in terms of species and age/size classes would vary over time, but, overall, the forest would be sustainable and self-perpetuating.

The description above is of what is generally understood as 'primary forest'. Large tracts of what were called climax forests and special forest formations have today been altered such that they are now degraded forests or secondary forests, and this trend is ongoing. As a result they have lower capacity to regenerate and reform their original composition other than over a huge time-scale. In cases where there have been major changes to the substrate and climatic pattern, such reformation will be effectively impossible. Their biodiversity composition will be much reduced as will the ecological processes that were delivered. At the same time, their long-term social and economic values will also be severely reduced.

Tropical forest management must adapt to the new paradigms observed during the past decades that have resulted in these changes. These include the rapid expansion of the agricultural frontier, over-exploitation of valuable or desirable species – not only tree species but perhaps heavy removal of a regeneration cohort for poles – climate change, invasive species and a rapidly increasing human population with progressively higher demand for products.

Preserving tropical forests by establishing strictly controlled protected areas has seldom worked effectively in tropical regions because of limited ability to apply sufficient resources to provide this and the very high relative value of at least some products. SFM aims to provide limited use of the products and services while not compromising the ability of the forest ecosystem to be sustained. This approach needs to deliver a flow of products and services that have sufficient value to act against pressures for conversion or devastation to provide a short-term windfall of returns that can never be repeated.

Restoring degraded natural forests is an option under what is today a major initiative in tropical areas, forest landscape restoration (FLR). The particular option of FLR can be typically implemented in tropical forest formations where socio-economic and environmental pressures have led to forest degradation, but not to complete destruction of forest cover and essential ecosystem services. Guided restoration can lead to forests that regenerate naturally and, combined with enrichment tree planting, help to protect ecosystems and regain productive capacities of forests. In production and protection forests alike, restoring natural forest formation will increase carbon storage, conserve biodiversity, increase watershed protection and enhance landscape resilience.

While it is inevitable that some forests will be lost through conversion, there are vast areas of degraded and secondary forests that can be restored and brought under sustainable management, including, where appropriate, development as highly productive plantations that can relieve the pressure of less productive natural forests. The challenge is to ensure that sufficient natural forest remains in a state that sustains their biodiversity and the complex web of interactions that supports this.

Sustainable forest management (SFM) must be predicated on a good understanding of the various forest types, their origin, composition, stand structure and dynamics that is to be managed. Complex tropical moist forests are most challenging, but a good knowledge of their regeneration system and care to maintain the complex structure together with its diverse assemblage of elements is crucial. Forest ecosystems that are naturally subjected to high levels of influence from agencies such as wind or fire are generally simpler and more robust than more complex ecosystems, but this does not mean they are insensitive or immune to these influences. Selective removal of certain desirable components from a forest ecosystem may lead to consequences far beyond the elimination of that component if its role within the overall system is not understood or respected.

To understand this challenging task, as well as to gain a good knowledge of forest ecology and taxonomy, it is increasingly important to recognise and work with the a wide range of stakeholders who have differential but important interests and requirements. The SFM of specific forest formations and forest types will need to be part of wider sustainable landscape management in which synergies are identified and captured, compromises made, and solutions developed, tested and revised as knowledge increases.

Regardless of whether it is complex natural forest or relatively simple plantation, maximum benefit will accrue by identifying, understanding and working with the natural processes that are going on within the forest ecosystem. This requires bespoke solutions developed from a good understanding of forest ecosystems and knowledge of the dynamics and interactions within them.

# 6  References

Aldrich, M., Bubb, P., Hostettler, S. and Van, H. (2000). *Tropical Montane Cloud Forests: A Time for Action, Arborvitae* (vol. 72). Available at: https://www.iucn.org/pt/content/tropical-montane-cloud-forests-time-action.

Alongi, D. M. (2014). Carbon cycling and storage in mangrove forests, *Annu. Rev. Mar. Sci.* 6, 195-219.

Blaser, J., Sarre, A., Poore, D. and Johnson, S. (2011). Status of tropical forest management 2011. *ITTO Technical Series No. 38*, Yokohama, Japan: International Tropical Timber Organization.

Brown, N. (1993). The implications of climate and gap microclimate for seedling growth conditions in a Bornean lowland rain forest, *J. Trop. Ecol.* 9(2), 153-168.

Cecil, D. J. and Blankenship, C. B. (2012). Toward a global climatology of severe hailstorms as estimated by satellite passive microwave imagers, *J. Clim.* 25(2), 687-703.

Chapman, J. D. and White, F. (1970). *Evergreen Forests of Malawi*, Oxford: Oxford University Press.

Corlett, R. T. (2016). Classifying tropical forests. In: Pancel, L. and Koehl, M. (Eds), *Tropical Forest Handbook*, Berlin: Springer.

Comeau, L., Hergoualc'h, K., Smith, J. U. and Verchot, L. (2014). Conversion of Intact Peat Swamp Forest to Oil Palm Plantation: Effects of Soil $CO_2$ Fluxes in Jambi,

Sumatra, Working Paper, Bogor, Indonesia: Center for International Forestry Research.

German, L. A. (2004). Ecological praxis and blackwater ecosystems: a case study From the Brazilian Amazon, *Hum. Ecol.* 32(6), 653–683.

Hahn, C. Z., Niklaus, P. A., Bruelheide, H., Michalski, S. G., Shi, M., Yang, X., Zeng, X., Fischer, M. and Durka, W. (2017). Opposing intraspecific vs. interspecific diversity effects on herbivory and growth in subtropical experimental tree assemblages, *J. Plant Ecol.* 10(1), 242–251.

Hamilton, S. E. and Casey, D. (2016). Creation of a high spatio-temporal resolution global database of continuous mangrove forest cover for the 21st century, *Glob. Ecol. Biogeogr.* 25(6), 729–738.

Hamilton, S. E. and Friess, D. A. (2018). Global carbon stocks and potential emissions due to mangrove deforestation from 2000 to 2012, *Nat. Clim. Change* 8(3), 240–244.

Hazard, C. (2018). Does genotypic and species diversity of mycorrhizal plants and fungi affect ecosystem function?, *New Phytol.* 220(4), 1122–1128.

Hollingsworth, L. T., Johnson, D., Sikaundi, G. and Siame, S. (2015). *Fire Management Assessment of Eastern Province, Zambia*, Washington, DC: USDA Forest Service, International Programs, 88 p.

Howe, H. F. (2014). Diversity storage: implications for tropical conservation and restoration, *Glob. Ecol. Conserv.* 2, 349–358.

Ifo, S. A., Moutsambote, J., Koubouana, F., Yoka, J., Ndzai, S. F., Bouetou-Kadilamio, L. N. O., Mampouya, H., Jourdain, C., Bocko, Y., Mantota, A. B., Mbemba, M., Mouanga-Sokath, D., Odende, R., Mondzali, L. R., Wenina, Y. E. M., Ouissika, B. C. and Joel, L. J. (2016). Tree species diversity, richness, and similarity in intact and degraded forest in the tropical rainforest of the Congo Basin: case of the forest of Likouala in the Republic of Congo, *Int. J. For. Res.* 2016, 1–12.

ITTO (2020). ITTO guidelines for forest landscape restoration in the tropics. *ITTO Policy Development Series No. 23*, Yokohama, Japan: International Tropical Timber Organization (ITTO).

Jacobs, M. R. (1979). *Eucalypts for Planting* (2nd edn.), Rome: FAO.

Lamprecht, H. (1990). *Silviculture in the Tropics*, Germany: TZ Verlagsgemeinschaft.

Kakishima, S., Morita, S. Y., Ishida, K., Hayashi, A., Asami, S., Ito, T., Miller, H., Uehara, D., Mori, T., Hasegawa, S., Matsuura, K., Kasuya, E. and Yoshimura, J. (2015). The contribution of seed dispersers to tree species diversity in tropical rainforests, *R. Soc.* 2(10), 150330.

Keith, H., Mackey, B. G. and Lindenmayer, D. B. (2009). Re-evaluation of forest biomass carbon stocks and lessons from the world's most carbon-dense forests, *Proc. Natl Acad. Sci. U. S. A.* 106(28), 11635–11640.

Malhi, Y., Phillips, O. L. L., Baker, J., Wright, T., Almeida, J. A., Arroyo, S., Frederiksen, L., Grace, T., Higuchi, J., Killeen, N., Laurance, T., Leaño, W. F., Lewis, C., Meir, S., Monteagudo, P., Neill, A., Núñez Vargas, D., Panfil, P., Patiño, S. N., Pitman, S., Quesada, N., Rudas-Ll, C. A., Salomão, A., Saleska, R., Silva, S. N. and Silveira, M. (2002). An international network to understand the biomass and dynamics of Amazonian Forests (RAINFOR), *J. Veg. Sci.* 13, 439–450.

Muchugi, A., Muluvi, G. M., Kindt, R., Kadu, C. A. C., Simons, A. J. and Jamnadass, R. H. (2008). Genetic structuring of important medicinal species of genus *Warburgia* as revealed by AFLP analysis, *Tree Genet. Genome* 4(4), 787–795.

Nguyen, H., Erfanifard, Y., Pham, V., Le, X., Bui, T. and Petritan, I. (2018). Spatial association and diversity of dominant tree species in tropical rainforest, Vietnam, *Forests* 9(10), 615.

Page, S. E. and Rieley, J. O. (1998). Tropical peatlands: a review of their natural resource functions with particular reference to South East Asia, *Int. Peat J.* 8, 95-106.

Page, S. E., Rieley, J. O., Shotyk, O. W. and Weiss, D. (1999). Interdependence of peat and vegetation in a tropical peat forest, *Philos. T. R. Soc.* 354, 1885-1897.

Page, S. E., Siegert, F., Rieley, J. O., Boehm, H. D., Jaya, A. and Limin, S. (2002). The amount of carbon released from peat and forest fires in Indonesia during 1997, *Nature* 420(6911), 61-65.

Page, S. E., Wust, R. A. J., Weiss, D., Rieley, J. O., Shotyk, W. and Limin, S. H. (2004). A record of Late Pleistocene and Holocene carbon accumulation and climate change from an equatorial peat bog (Kalimantan, Indonesia): implications for past, present and future carbon dynamics, *J. Quat. Sci.* 19(7), 625-635.

Prideaux, B. (2014). *Rainforest Tourism, Conservation and Management: Challenges for Sustainable Development*, Routledge: The Earthscan Forest Library.

Pryor, L. D. (1976). *The Biology of Eucalypts*, London: Edward Arnold.

Roberts, P., Boivin, N., Lee-Thorp, J., Petraglia, M. and Stock, J. (2016). Tropical forests and the genus Homo, *Evol. Anthropol.* 25(6), 306-317.

Salati, E., Marques, J. and Molion, I. C. B. (1978). Origin and distribution of precipitation in Amazonia, *Interciencia* 3, 200-205.

Sedio, B. E., Rojas Echeverri, J. C., Boya P, C. A. and Wright, S. J. (2017). Sources of variation in foliar secondary chemistry in a tropical forest tree community, *Ecology* 98(3), 616-623.

Spalding, M. D., Blasco, F. and Field, C. D. (1997). *World Mangrove Atlas*, Okinawa: The International Society for Mangrove Ecosystems.

Spalding, M., Kainuma, M. and Collins, L. (2010). *World Atlas of Mangroove*, London and Washington: Earthscan Press.

Stadtmueller, T. (1987). *Cloud Forests in the Humid Tropics: A Bibliographic Review*, Turrialba: CATIE.

Tarakeswara, M., Premavani, D., Suthari, S. and Venkaiah, M. (2018). Assessment of tree diversity in tropical deciduous forests of Northcentral Eastern Ghats, India, *Geol. Ecol. Landscapes* 2(3), 216-227.

Ter Steege, H., Pitman, N., Sabatier, D., Castellanos, H., Van Der Hout, P., Daly, D. C., Silveira, M., Phillips, O., Vasquez, R., Van Andel, T., Duivenvoorden, J., De Oliveira, A. A., Ek, R., Lilwah, R., Thomas, R., Van Essen, J., Baider, C., Maas, P., Mori, S., Terborgh, J., NúÑez Vargas, P., Mogollón, H. and Morawetz, W. (2003). A spatial model of tree ⊡-diversity and tree density for the Amazon, *Biodivers. Conserv.* 12(11), 2255-2277.

Thomas, S. C. and Baltzer, J. L. (2002). *Tropical Forests. Encyclopedia of Life Sciences*, London: Nature Publishing Group.

Tovo, A., Suweis, S., Formentin, M., Favretti, M., Banavar, J. R., Azaele, S. and Maritan, A. (2016). *Biodiversity of Tropical Forests*, Cold Spring Harbor Laboratory, 1-21.

Troup, R. S. (1932). *Exotic forest trees in the British Empire*, Clarendon Press, Oxford, UK.

Umaña, M. N., Forero-Montaña, J., Muscarella, R., Nytch, C. J., Thompson, J., Uriarte, M., Zimmerman, J. and Swenson, N. G. (2016). Interspecific functional convergence and divergence and intraspecific negative density dependence underlie the seed-to-seedling transition in tropical trees, *Am. Nat.* 187(1), 99-109.

Weischet, W. (1980). *Die ökologische Benachteiligung der Tropen*, Stuttgart: Teubner Verlag.

Whitmore, T. (1998). *An Introduction to Tropical Rain Forests*, Oxford: Oxford University Press.

Wingfield, M. J. and. Swart, W. J. (1994). Integrated management of forest tree diseases in South Africa, *For. Ecol. Manag.* 65(1), 11–16.

WRI (2015). Satellite data reveals state of the World's Mangrove Forests, Available at: https://www.wri.org/blog/2015/02/satellite-data-reveals-state-world-s-mangrove-f orests (Accessed 1 April 2020).

Wright, J. S. (2002). Plant diversity in tropical forests: a review of mechanisms of species coexistence, *Oecologia* 130(1), 1–14.

Zakaria, M., Rajpar, M. N., Ozdemir, I. and Rosli, Z. (2016). Fauna diversity in tropical rainforest: threats from land-use change. In: *Tropical Forests: the Challenges of Maintaining Ecosystem Services While Managing the Landscape*, IntechOpen.

# Chapter 2

## Defining sustainable forest management (SFM) in the tropics

*Francis E. Putz, University of Florida-Gainesville, USA; and Ian D. Thompson, Thompson Forest Ltd.-Kelowna, Canada*

## 1  Introduction

> 'Any claim of sustainable forest management should evoke the queries: What is sustained? What were the tradeoffs? Over what spatial and temporal scales?'

Sustainable forest management (SFM) is a conceptual codification of forest management practices that continues to evolve from its focus in the 1800s on sustained timber yields. Since the 1987 publication of 'Our Common Future' (also known as the Brundtland Report) by the World Commission on Environment and Development, the definition of SFM has expanded to include the much broader goals of sustaining the economic, social, and environmental benefits from forest. In the words of the United Nations, SFM is a 'dynamic and evolving concept [that] aims to maintain and enhance the economic, social, and environmental values of all types of forests, for the benefit of present and future generations' (FAO, 2018). This broadening of considerations is reflected in the definitions provided by the International Timber Trade Organization (ITTO):

http://dx.doi.org/10.19103/AS.2020.0074.19

> [SFM is] the process of managing forest to achieve one or more clearly specified objectives of management with regard to the production of a continuous flow of desired forest products and services without undue reduction of its inherent values and future productivity and without undue undesirable effects on the physical and social environment. (ITTO, 2016)

In this chapter, we elaborate on these definitions of SFM in an effort to promote clarity about the avoidable and unavoidable trade-offs associated with all management decisions; management 'for' something is necessarily management 'against' something else. We also hope to promote measurement, monitoring, and verification of the various indicators of sustainability. It is written out of concern for the obvious deniability of many claims of SFM and the inclusion of vague terms in its definitions such as 'over the long term' without specification of time scales and 'without undue reduction' without clarification of 'undue'. In the chapter we also advocate for clarity about spatial scales and for expansion of the scale at which sustainability is considered from stands up to forested landscapes. Finally, we believe that graphical depictions of the components of SFM, like the one we propose, will help clarify these trade-offs and generally aid our understanding about the challenges of reaching the SFM goal.

Our approach follows the effort of Thompson et al. (2013) to clarify the complex condition of 'forest degradation' through its disaggregation into component biophysical parts. We hope that our un-clustering of the various dimensions of SFM will similarly help inform efforts to promote and evaluate forest management sustainability. We also expand the scope of SFM from individual stands, to which many definitions of SFM pertain, to the scale of forested landscapes, in keeping with other efforts towards the comprehensiveness of land use planning (e.g. Sayer et al., 2016). We hope that our efforts are of use in the development of principles, criteria, and indicators of sustainability for programs such as the Sustainable Landscape Production Certification program under development by the Landscape Standard Consortium (https://verra.org/project/landscape-standard/).

We proceed in this effort to clarify landscape-scale SFM by defining its principal components and then considering them at different spatial and temporal scales. We strive for measurability and precision, in recognition of the diversities of landscapes with forests, characteristics of managed forests and forest managers, forest management goals, and trade-offs associated with land-use interventions. We nevertheless recognize that any definition of SFM with wide applicability and acceptability must be somewhat vague and mutable. That said, whatever the definition of SFM that is adopted, clarity and measurability should be fundamental objectives.

## 2 Evolving concepts of sustainability

Since well before the Brundtland Report (1987), economists have grappled with what is meant by 'sustainable' and 'sustainability' (e.g. Solow, 1956). While the concerns of foresters about sustainability date back many centuries (reviewed by Wiersum, 1995), the focus was historically on sustaining timber yields, with non-diminishing yields being the goal of management. This focus, which remains relevant, is now referred to as 'strong' sustainability (reviewed by Luckert and Williamson, 2005). In contrast, 'weak' sustainability allows for the transfer of natural capital (e.g. timber stocks or biodiversity) for economic, built, social, and human capital as long as the overall sum of these five forms of capital does not decline. Recognition of the embeddedness of managed forests in landscapes of various other forest and non-forest land uses is more recent, and is reflected in what are known as landscape-level and jurisdictional approaches to sustainability (e.g. Sayer et al., 2016; Stickler et al., 2018; Runting et al., 2019; Griscom et al., 2019.

Expansions of SFM's scope were unavoidably accompanied by modification of the definition of 'sustainability' from one that requires non-diminishing supplies to one that is much more multi-dimensional and negotiable. One consequence of this broadening of the definition and the allowance for capital transfers is that it allows claims of 'sustainable development' and 'sustainable infrastructure'. The expansion of the concept of 'sustainability' to non-renewable resources, as exemplified by the *Journal of Sustainable Mining*, suggests that 'sustainable' is now just a synonym for 'responsible' or 'good' (Putz, 2018).

Here we consider a definition for SFM in the realm of tropical forests that accounts for multiple classes of managed, exploited, and unmanaged forests across landscapes that can include protected areas, selectively logged natural forests, logged forests subjected to additional silvicultural treatments to increase stocking and growth of commercial species, plantations, and forest restoration areas (Fig. 1). We also separate out for consideration forests under the control of rural, local, and/or indigenous communities in full recognition that their lands may host any of these sorts of management practices. Our approach to SFM differs from multiple-use forest management, which typically focuses on compromising goals in areas subjected to similar treatments in what has become recognized as 'land-sharing' (e.g. Phalan et al., 2011). Our approach also expands the 'triad' concept of Messier et al. (2009) in which the focus is on natural forest management, plantation forestry, and forest protection by additionally considering community-based forest management and forest restoration. We hope to shed light on the various benefits derived from different portions of landscapes with forests. We strongly recommend that this approach be expanded by the inclusion

**Division of forest landscape for SFM:**

1. Protected areas ▮
   - variables: size, location, connectedness, % area
2. Extensive forest management ☐
   - variables: values to achieve sustainably, % area
3. Plantation forest management ▦
   - variables: size of plantations, species, purpose, % area
4. Community forest management ▨
   - variables: values important to community, % area
5. Forest restoration ▩
   - variables: values to achieve sustainability; % area

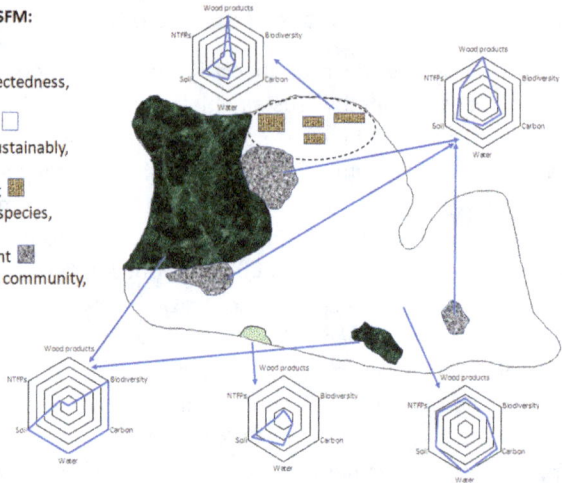

**Figure 1** Partitioning a forest landscape for assessment of SFM (extensive and intensive natural forest management not differentiated, but the latter should occur in accessible areas such as near main roads). Each category of forest land-use is evaluated on the basis of the same six criteria illustrated by the biophysical resource hexagons. The overall sum of scores is a relative measure of SFM at the landscape scale for a particular year (The concentric lines inside the perimeter of the main polygons refer to % values of the indicators relative to a primary forest baseline; the blue lines are examples of monitored values from one year).

of more values that are social and economic, and consideration of other land-uses.

The criteria on which we focus are wood products, non-timber forest products, soils, water, carbon, and biodiversity. We assume that the aggregate measure of the extent to which these values are maintained is a measure of the degree of SFM. We recognize that this approach to the assessment of SFM is mostly restricted to biophysical attributes affected by intentional forest management and forest resource exploitation, but we do consider the sustainability of profits from the sale of timber and non-timber forest products (see below). Social, cultural, and other sorts of criteria for more complete assessments of SFM should be readily added to this basic system. Another possible modification that deserves consideration is the use of asymmetrical polygons that illustrate differences in emphases on the various values/criteria.

## 3  Appropriate scales for assessment of SFM

One difficulty with the concept of SFM is the uncertainty about the scale at which it should be assessed. While forestry practices are implemented at stand scales, given the many values of forests and the inherent trade-offs in any stand-level management regime, SFM might more logically be considered at landscape scales (e.g. Vincent and Binkley, 1993; Boscolo, 2000). In other words, stand-scale management cannot maintain everything, everywhere, all the time, nor should it aim to do so. Only at the landscape level can all forest values be sustained over time and space, if managed properly. With our approach, the sum of scores on all the objectives (i.e. axes in the trade-off polygons), weighted by the area of each land-use, represents a landscape-level measure of sustainability for all criteria (i.e. goods, services, water, carbon, biodiversity, recreation, etc.) at one point in time. We suggest that, with the exception of the most intensive short-rotation tree plantations, multiple-use management with multiple hoped-for benefits is likely to be the goal for most portions of managed forest landscapes. We also recognize that the constraints on achieving the goal of multiple forest management are basically the same, and equally as daunting, as those for SFM in the tropics (Sabogal et al., 2013).

Given the diversity of forest management options, each with its own inherent trade-offs, as well as the diversity of forest conditions, a landscape approach seems appropriate as a first step toward figuring out how the undesired outcomes of management can be minimized overall. Explicit recognition of trade-offs among land uses allows increased rationality in their assessment, as opposed to attempts to maintain all values everywhere all the time. Sizes of managed landscapes sufficient for SFM are likely to be dictated by existing natural constraints, negotiation, geography, and politics, but local values may bound all the others at the upper end of the spatial scale. In these cases, landscape size should reflect some value that would unavoidably be depleted or its maintenance rendered uneconomical as a result of managing at too small a scale. For example, rare species of trees or large mammals may require several thousand square kilometers for persistent populations (e.g. Schulze et al., 2005; Wikramanayake et al., 2011). In these cases, we do not suggest managing for the minimum viable population, but rather some upper value that accounts for temporal stochasticity as well as for controlled and uncontrolled exploitation. In other geographies, for SFM to be practicable, a sufficiently large area may be required to provide an adequate economic benefit from a valued resource (Nasi and Frost, 2009; Sabogal et al., 2013).

## 4   SFM trade-offs at different scales

Management, by definition, requires that when some species, conditions, processes, or values are managed for, some other species, conditions, processes, or values are managed against. In other words, trade-offs are as inherent to the act of management as they are to resource exploitation. In tropical forests, SFM requires consideration of sufficiently large landscapes for all values to persist, enabling sustainable wood production, sufficient ecosystem services for communities, and no losses of species. For this purpose, areas within the landscape used for different purposes are partitioned into use-categories, often with different values or benefits to society (Fig. 1). For example, intensively managed plantations have limited value for biodiversity but high value for commercial wood production, while protected areas are the opposite. These differences in value-maintenance among land-use categories are represented by the blue value lines inside the trade-off polygons; value lines that approach the outer perimeter of the polygon represent value maintenance relative to the primary forest baseline. Indicators for each of the six criteria are landscape-specific and depend on the specific circumstances and the selected objectives for each forest category. For example, a key indicator for protected areas might be elephant population persistence, while for a plantation, indicators might be a certain amount of wood, fuel, or rubber produced per year. We illustrate five forest categories (Fig. 1), but there are others, such as local community conservation areas and private lands, that may deserve consideration. Regardless of the number of forest categories, each can be evaluated on the basis of the same six criteria in full recognition of the different objectives for which different land-use categories are managed. For more complete assessments of SFM, criteria need to be added that capture the social, economic, and additional environmental values. Extension of this approach to non-forest land uses is possible, but will likely involve specification of new evaluation criteria.

## 5   Defining terms in SFM

To clarify how SFM might be attained and measured, we commence with definitions of forest (as opposed to plantation), management, and sustained yield. Again, we do not believe that our definitions are sacrosanct, but argue that agreed-upon definitions are needed lest discussions of SFM continue to be plagued by vagueness and ambiguity. A limitation of our approach is that the focus is on forest landscapes and excludes land cleared for agriculture, mining, impoundments or other non-forest land uses, even those with substantial tree cover (e.g., urban forests and some agroforestry systems). We also recognize that our focus is principally biophysical, but hope that our approach is sufficiently adaptable to accommodate social and economic considerations.

*Forest versus Plantation*: Tree-covered landscapes among which forests with natural regeneration are differentiated from plantations in which all future crop trees are planted (Sasaki and Putz, 2009; Putz and Redford, 2010) often with the intention to clear-cut at frequent intervals. This distinction is made in full recognition of intermediate states, such as selectively logged natural forests that are enriched by planting along cleared lines or in felling gaps. We also recognize that the deleterious environmental impacts of intensive plantation management can be mitigated in many ways such as by maintaining natural forest corridors in riparian areas, increasing structural and floristic diversity within stands, and extending rotations (e.g. Dudley, 2005).

*Management*: Intentional actions are taken with specified goals, to differentiate management from exploitation and its consequence, degradation. Management occurs at multiple scales and intensities, as fitting for different objectives and in recognition of different trade-offs, and includes protection as well as harvesting. For natural forest management, whether conducted by communities or industrial firms, we differentiate low intensity but extensive approaches based on reduced-impact logging, from higher intensity sorts of silvicultural interventions (e.g. liberation thinning and enrichment planting).

*Sustained Yields*: The topic of sustained yield forestry has received attention for centuries and may seem more straightforward a consideration than biodiversity, aesthetics, or other values, but we believe it is helpful to disaggregate claims of sustained yield for assessment purposes (Fig. 2). Although most of the data about the effects of sequential harvests are for timber, we believe the same situation applies to non-timber products, especially those for which individuals are harvested in their entirety (e.g. rattan palms).

In the more in-depth assessment of sustained yield proposed here, the degree to which volumetric yields are maintained from one harvest to the next is retained but only as one criterion, described by one axis in a trade-off pentagon (Fig. 2). Given the propensity for harvesters to 'high-grade' (i.e. to select the best individuals first), product quality typically declines with each successive harvest (e.g. increased prevalence of crooked, small, hollow, and heart-rotted trees); we capture this trend in another axis in the sustained yield pentagon. Included in this dimension of sustained yield would be changes in wood densities and working properties such as between old-growth timber and that of regenerating stands of fast-growing trees. The similar tendency to harvest the biggest trees first is represented by an axis that reflects the

## Sustained Yield

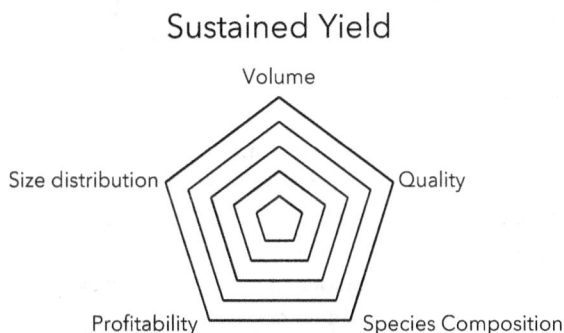

**Figure 2** Sustained yield assessed by five criteria. The concentric lines refer to % values of the indicators relative to primary forest (outermost line).

size-class frequency distributions of stems in the post-harvest forest, at a point in time. Given the disproportionately large contributions of very large trees to forest structure, biodiversity maintenance, ecosystem processes, and both population and carbon dynamics (e.g. Lindenmayer et al., 2012; Slik et al., 2013; Sist et al., 2014; Thompson et al., 2014; Kohl et al., 2017), their retention in managed forests is of substantial environmental importance. This importance is reflected in the Brazilian forestry regulation that requires retention of at least 15% of all large trees or the three largest trees per 100 ha harvest block (CONAMA, Resolution no. 1 of 2015; Vidal et al., 2020). Also included in the yield component of SFM is an axis that reflects the sustainability of financial profits. As with all the trade-off polygons, a composite value for sustained yield is calculated as the sum of these five components. Other considerations might, of course, be included and the components might be differentially weighted, but the overall approach provides some clarity about the assessment of claims of sustained yields.

## 6  Land-use types in SFM

To secure the benefits of landscape-level assessments of SFM, landscapes need to be subdivided into different land-use categories. For a theoretical forested landscape in the tropics, we here consider the following six land-use types:

- 1 Protected Areas;
- 2A Natural Forest Management with Selective Harvests of Timber and Non-Timber Forest Products;
- 2B Natural Forest Management with Silvicultural Treatments After Selective Logging;
- 3 Tree Plantations;
- 4 Community Forests; and
- 5 Forest Restoration Areas.

### 6.1 Protected areas

Protected areas are designated to maintain ecosystem processes, protect biodiversity and especially low-density species, maintain high carbon stocks, and provide ecosystem services to surrounding landscapes and people. For an example of their importance in a landscape context, one of the few studies that found convergence of conditions in managed forests to those in primary forest noted the importance of associated large protected areas to species and ecosystem process recovery (Norden et al., 2009).

Intact ecosystems in protected primary forests also provide benchmarks against which to measure SFM. Individual protected areas are often not large enough on their own to protect wide-ranging or low-density species. At the landscape scale, proper management of surrounding buffer areas can provide the connectivity among protected areas that is required for persistence of some species (e.g. Hodgson et al., 2011). Like SFM in general, the effectiveness of tropical protected areas is very much a function of governance, stakeholder agreement, level of staff training and commitment, and sufficient funding (Bruner et al., 2001). Assessments of the extent to which protected areas deliver the expected or hoped-for values are made challenging by the tremendous variation in the degree to which protected areas are essentially abandoned or are actively protected with controls on access and resource exploitation.

Unmanaged or primary forests (i.e. those with no visible signs of human intrusion; FAO, 2018) are declining rapidly, especially those that are large. Potapov et al. (2017) reported that globally, the area of intact forest, defined as areas of >500 km$^2$ with no roads, declined by 7.2% between 2000 and 2013; such areas are already absent in many tropical countries. It is abundantly clear that the stocks of carbon and biodiversity in large primary forests exceed those in forested lands subjected to uses other than protection (e.g. Barlow et al., 2007; Luyssaert et al., 2008; Pan et al., 2011; Edwards et al., 2014; Watson et al., 2018). Many large-bodied and/or heavily exploited tropical animal species prefer intact forests, including the Asian elephant (*Elephas maximus*), African forest elephant (*Loxodonta cyclotis*), tiger (*Panthera tigris tigris*), and harpy eagle (*Harpia harpyja*) (Kinnaird et al., 2003; Barnes et al., 1991; Barlow et al., 2011; Birdlife International, 2016, but see Roopsind et al., 2017). Complicating the discussion of the conservation value of intact forest is the research demonstrating that up to 94% of the area in blocks of forest designated for selective logging remained intact due to the absence of commercial timber, adverse conditions, or poor planning and inadequate supervision (mean = 69%; Putz et al., 2019).

The vast majority of biodiversity exists outside protected areas and the ranges of many species protected partially or predominantly inside parks extend well beyond park boundaries. Hence, protected areas can rarely maintain viable populations of low-density species. Furthermore, in many areas of the world, protected areas either do not exist (Rodrigues et al., 2004) or are unmanaged and subject to illegal activities including poaching and logging (Loveridge et al., 2007; Wittemyer et al., 2008). Laurance et al. (2012) suggested that at least half of Earth's protected areas are failing to sustain their biodiversity. While the rate of loss of intact forests has generally been higher outside areas designated for protection, intact forest areas inside parks nevertheless often decline. For example, Virunga National Park in the Republic of Congo lost 3.3% of its forest cover in just over a decade (Potapov et al., 2017). Overall, the extensively managed forests that serve as buffers for protected areas are essential to sustain tropical biodiversity.

## 6.2 Natural forest management with selective harvests of timber and non-timber forest products

Natural forests managed for timber and non-timber forest products, which in the tropics typically involves selective harvests, maintain many values when managed properly, as detailed below (Fig. 3). Unfortunately, despite substantial expenditures of time, money, and effort, yields from the harvested species are seldom maintained even when governmental regulations are scrupulously followed (Putz et al., 2012; Vidal et al., 2020). Generally the most valuable timber species are harvested first, followed successively by each of the less valuable ones in subsequent harvests, often referred to as 'logging down the value chain' (e.g. Schaafsma et al., 2013). Furthermore, most government agencies and non-governmental certification bodies (e.g. the Forest Stewardship Council) lack the wherewithal to determine if yields from individual species or even entire forests are maintained (Romero and Putz, 2018).

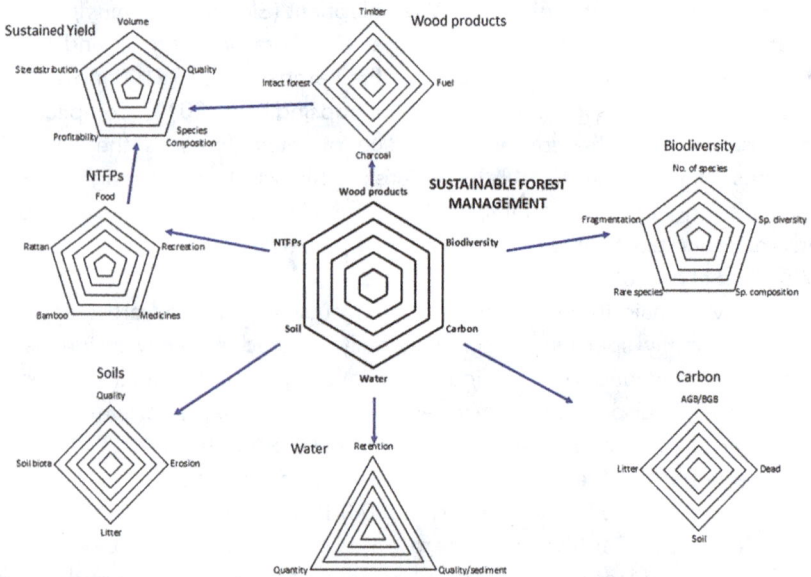

**Figure 3** A suggested approach to disaggregation of SFM into its component values with emphasis on biophysical attributes. Note that the criteria and indicators for evaluation of sustained yield pertains to both timber and non-timber forest products. The concentric lines refer to % values of the indicators relative to a primary forest baseline.

Under current regulations in most tropical countries, timber stocks do not regain primary-forest volumes by the end of each officially designated minimum cutting cycle. After conventional timber harvests in Amazonian Brazil, for example, timber volumes take >60 years to recover, not the 25–30 years allowed by law (reviewed by Vidal et al., 2020). Some studies report eventual convergence of logged forests to primary forest conditions (e.g. Norden et al., 2009), while others suggest convergence will not occur (de Avila et al., 2015). A meta-analysis of studies on yield recovery based on >100 publications (Putz et al., 2012) revealed substantial variability, but concluded that timber yields declined by about 46% from the first harvest to the second harvest. That study also reported that, on average, 76% of carbon is retained in forests logged once, and that 85–100% of species of mammals, birds, invertebrates, and plants remain after logging, although long-term persistence is not assured. It is important to note, however, that such studies only report on a few taxa and do not consider all ecosystem functions, especially those delivered by complexes of co-evolved species. Furthermore, the studied forests were not selected at random and were likely representative of the best management underway when the studies were conducted. Finally, although many tropical forests are being logged for the second or third time, most of the reviewed studies focused on timber harvests from primary forest. One consistent message is that, despite the conservation potential of extensive selective logging, SFM is currently jeopardized in much of the tropics by poor logging practices (e.g. Ellis et al., 2019) and premature re-entry logging of previously harvested stands (Sasaki et al., 2016).

The value of extensive production forests for biodiversity and ecosystem services varies with logging intensity (e.g. Burivalova et al., 2014; Franca et al., 2017), logging practices (e.g. Pinard and Putz, 1996; Vidal et al., 2016), but particularly with post-harvest secondary effects including deforestation, poaching, and illegal logging (Michalski and Peres, 2013; Zimmerman and Kormos, 2012; Specht et al., 2015). It is clear that if logging and access are controlled, these secondary effects can be avoided and extensive areas of selectively logged forest will maintain considerable conservation value (Edwards et al., 2011; Putz et al., 2012; Edwards et al., 2014; Lewis et al., 2015; Roopsind et al., 2018, but see Laufer et al., 2013). While extensively managed forests can support much biodiversity, selective logging may nevertheless have substantial deleterious impacts on populations of high-value tree species and associated fauna (e.g. Fisher et al., 2011), partially due to losses of seed sources and dispersal agents. These populations can often be recovered only through carefully managed planting (see 2B).

The ITTO suggested that at least 500 million ha of tropical forest were degraded by 2002; we are not aware of any more recent global estimates of degradation except specifically for carbon (e.g. Baccini et al., 2017). While global attention has swung to reforestation of tree-free areas, restoration of these degraded forests should be a priority. Restoration of degraded forests implies increasing forest resilience, reducing the probability of successful invasions by exotic species, emulating natural processes in silvicultural regimes, and especially avoiding continued degradation. For instance, managing to ensure resilience means maintaining natural species composition and the capacity of forest composition to change under natural circumstances, with only gradual shifts in structure and function (Thompson et al., 2009). Accomplishment of this objective requires that forests are not degraded to a tipping point beyond which the ecosystem state changes radically to a novel and potentially stable condition (e.g. closed forest to open forest). Forest degradation is a difficult concept, however, owing to different perceptions for values derived from the forest, but generally refers to the loss of goods and services (CPF, 2010; Vásquez-Grandón et al., 2018). Degradation becomes easier to measure when considered at the landscape scale by using criteria and indicators (Thompson et al., 2013), like those proposed herein for SFM (Fig. 3) For example, many observers consider plantation to be highly degraded forests or not forests at all (Putz and Redford, 2010), to distinguish them from natural forest. Where plantations replace forests, the environmental losses cannot be recovered, but where plantations are established in already deforested areas, make up only a small proportion of the landscape, are societally accepted, and reduce pressure on natural forests as sources of wood products, they may be acceptable on both environmental and economic grounds. There are also many ways that the deleterious impacts of plantations can be mitigated (e.g. Dudley, 2005), but, based on our observations, few of the recommended practices are ever implemented at industrial scales.

The prerequisite conditions for SFM include proper policy support and legal frameworks, sufficient worker training, uncontested land tenure, sufficient financial incentives, and effective enforcement of regulations (e.g. Nasi et al., 2011; Sabogal et al., 2013; ITTO, 2015, 2016). Lack of financial remuneration for the many environmental services provided by natural tropical forests is one reason for the low financial competitiveness of forest management compared to other land uses such as agriculture and cattle-ranching. To the extent that reduced-impact logging is synonymous with reduced-income logging, it is not reasonable to expect loggers to adopt improved harvesting practices out of enlightened self-interest (Putz et al., 2000). Payments for ecosystem services (PES) seem like a viable mechanism to promote SFM, but successful uses of this tool are scarce and the benefits are ephemeral and funding dependent. It is noteworthy that the 'Socio Bosque' PES program in Amazonian Ecuador reportedly promoted reductions in both deforestation and forest degradation (Mohebalian and Aguilar, 2018). Aside from financial incentives, strong enforcement is also essential to secure the benefits from the vast quantities of carbon that could be sequestered by improved forest management (e.g. Pinard and Putz, 1996; Vidal et al., 2016; Ellis et al., 2019). Halting land grabs and poaching of both wildlife and timber is rendered especially difficult after logging roads improve access and thereby the profits from illegal activities, but the benefits from enforcement are substantial (e.g. Roopsind et al., 2018). Carbon crediting from improved management is sanctioned by the United Nations' Reduced Emissions from Deforestation and Forest Degradation (REDD+) program, but funds remain scarce for promoting the transition from forest exploitation to forest management.

## 6.3 Natural forest management with silvicultural treatments after selective logging

The principal intervention in tropical forests designated for timber production is selective logging. If properly conducted, selective logging can be considered as a silvicultural technique insofar as it can promote the regeneration and growth of commercial species (e.g. Vidal et al., 2016). Unfortunately, despite decades of promotion of reduced-impact logging (RIL) including millions of dollars spent on RIL policy development and training, most logging still more closely represents timber mining than timber stand management (Ellis et al., 2019). Foresters concerned about logging-induced reductions in timber yields and timber quality, as well as the sequential extirpation of commercial species with each harvest have long prescribed silvicultural interventions. Research firmly establishes the silvicultural benefits of these treatments, but apparently due to insufficient motivation, they are seldom applied outside of research plots.

The toolbox for tropical silviculture includes interventions that range from variations on felling regimes (e.g. strip clear-cuts and group selection harvests), pre-felling treatments such as the cutting of lianas on trees to be felled, as well as the planning of extraction pathways and the marking of trees for directional felling (i.e. RIL). Post-logging silvicultural treatments include liana cutting on future crop trees (FCTs), liberation of FCTs from arboreal competitors, mechanical scarification of felling gaps to promote regeneration, culling of non-commercial trees, and enrichment planting of commercial species along cleared lines or in felling gaps.

High harvest intensities in natural forests typically remove the valuable, mostly shade-tolerant hardwoods, while it damages young recruits, which leads to non-recovery of these species (Van Gardingen et al., 2003; Anitha et al., 2010). Often even low-intensity harvesting can deplete the valuable species (Peña-Claros et al., 2008; Sebben et al., 2008; Schulze et al., 2008a,b, Kukkonen and Hohnwald, 2009), hence the necessity of assisting natural regeneration of over-exploited species, such as with enrichment planting of mahoganies (*Swetenia* spp.), rosewoods (*Dalbergia* spp.), ipê (*Tabebuia* spp.), and cedar (*Cedrela* spp.). The silvicultural effectiveness of each of these treatments is supported by research, but even for these high-value species, few are applied outside of research areas (but see Navarro-Martínez et al., 2017). Historically, more broad-scale employment of silviculture occurred, such as the application of the Malayan Uniform System in Malaysia, but those treated stands were mostly converted to oil palm plantations and silvicultural treatments were discontinued in the forests that remained forest.

For reasons that are not completely clear but that include improved governance and increased recognition of current and pending shortfalls of timber supplies, there are now a few commercial-scale examples of silvicultural intensification of natural forest management (Puettmann et al., 2015). For one, liana cutting on future crop trees (i.e. trees smaller than the minimum cutting diameter that are expected to mature by the end of the cutting cycle) is reportedly more the rule than the exception in a major logging concession in Belize (Mills et al., 2019). Another example is in Indonesia where at least one logging concession carries out large-scale enrichment planting along cleared lines through twice-logged forest (Ruslandi et al., 2017a).

## 6.4 Tree plantations

While the area of primary forest in the tropics declines, the area of tree plantations (which we do not consider forest *sensu latu* because of the limited species composition, rapid turnover, and usually single objective for wood fiber; Putz and Redford, 2010) increased dramatically over the past two decades. Planted forests now cover >278 million hectares, increasing from 4% to 7% of the reported total tree-covered area between 2010 and 2015 (Payn et al., 2015). Here, we differentiate assisted natural regeneration of native species in extensively managed natural forests, involving directed post-harvest silvicultural treatments, from intensive plantation forestry. Among the plantations are those under short-term (i.e. fastwood) and longer-term cutting cycles, but most often involve a single species used for utility grade timber, chips and fibers, or fuel (Brockerhoff et al., 2008) and support limited biodiversity. Commonly planted are species of the genera *Acacia*, *Eucalyptus*, and *Pinus*. Our justification for this distinction is that forests with assisted natural regeneration also contain many naturally recruited trees and the planted species would not have recovered naturally without the intervention (Thompson et al., 2014; Ruslandi et al., 2017b). We note that although the majority of plantations we have observed in the tropics are appropriately considered 'green deserts', there is plenty of research demonstrating the benefits of biodiversity-enhancing design and management practices such as mixed species plantings and retention of natural forest along riparian corridors (e.g. Dudley, 2005; Paquette and Messier, 2010; Liu et al., 2018). We also note that the assumption that plantations take the pressure of natural forests (e.g. Sedjo and Botkin, 1997) seems supported under some conditions, but remains to be rigorously tested.

## 6.5 Community forests

We include community forests as a separate land-use category because, although they may be subjected to many different management practices, we assume that SFM is a principal goal. In some cases, however, community forests can also be fully protected, used for ecotourism purposes, or managed by commercial contractors. In any case, many tropical countries are trying to reduce the deleterious impacts of concession forestry and to redress prior local communities' rights violations by assigning management responsibilities to these constituencies. The ethical appropriateness of retuning land to traditional owners notwithstanding, the impacts of community forestry range from relative successes insofar as management improved to failures with rates of deforestation that do not differ from other forests (Bowler et al., 2012; Santika et al., 2017). These failures reportedly resulted from a combination of a lack of training, insufficient funding, disinterest by government in reviewing progress, lack of agreement and coherence of action among community members, land-grabbing, and various illegal/informal activities. In contrast, Porter-Bolland et al. (2012) found that 33 community forests generally had lower rates of deforestation than 40 protected areas, but the mechanisms responsible for this environmental benefit could not be specified due to lack of clear counterfactuals (i.e. what would have happened in the absence of community land tenure). Furthermore, if over time communities accumulate capital and increase in market integration, land-use practices may intensify especially if they are allowed to sell or lease their land.

## 6.6 Forest restoration areas

Given global attention to the potential benefits of forest restoration (e.g. Griscom et al., 2017), we include this land use, but with some misgivings. One cause of concern is that many of its proponents fail to distinguish between plantations and forests, so that the result of reforestation interventions can differ fundamentally. It is also often unclear whether forest products can be harvested from the reforested areas. Some projects do aim for full ecological restoration, which means recovering the species diversity and composition of primary forest, but it is not clear that this ambitious goal is attainable. Finally, differences in starting conditions affect the outcomes of restoration interventions. For example, the likely outcomes of forest restoration differ between areas that were deforested and then plowed, planted, fertilized, or overgrazed from those that suffered only a clear-cut. Spatial scale and landscape settings also matter, especially if propagules for regeneration need to be dispersed to great distances. In any case, restoration efforts are generally new and of limited overall consequence for the landscape, at least by the year 2020. Worst of all, when naturally tree-poor savannas and grasslands are afforested, the biodiversity consequences are grave (e.g. Veldman et al., 2015).

## 7   Challenges for SFM in the tropics

To various extents, tropical landscapes present a special case for the implementation of SFM that reinforces the need for disaggregated approaches to assessment, like the one proposed herein. First of all, many forested areas in the tropics are characterized by weak governance, contested land ownership, poverty, large numbers of forest-dependent people, rapid rates of exploitation and forest conversion, modest-to-high opportunity costs of forest retention, and/or political conflicts.

Considerations of sustainability are further complicated by the fact that far more wood is taken for fuel than for timber. For example, Sprecht et al. (2015) found that annual demand for fuelwood by 210 municipalities in Amazonian Brazil was about 300 thousand tons, which they noted would require the clearing of 1200–2100 ha of forest.

Efforts at SFM often face challenges related to the legacies of former interventions: many of the forests exploited for timber today were previously logged, either legally or illegally, but virtually always with little regard for the future. Even in forests with no recent history of exploitation, given high species diversity, tropical trees that produce commercial timber are generally scarce and patchily distributed, which can lead to their rapid commercial extirpation. Many such species, including rosewoods and mahoganies, are now listed by CITES (Convention on International Trade in Endangered Species of Wild Fauna and Flora). Growth and regeneration rates are such that, to maintain viable populations of many commercial timber species, large areas and low-harvest rates are required, at least unless silvicultural interventions, such as liberation thinning around future crop trees and enrichment planting, are applied. Complicating matters further is the fact that some of these species require interior forest conditions and are light, moisture, and thermally sensitive. Furthermore, many tree species depend on co-evolved relationships for pollination, seed dispersal, and nutrient acquisition (e.g. Lewis, 2009; Campos-Arceiz and Blake, 2011).

Given the widespread conversion of lowland forest on gentle terrain to more intensive land uses, natural forest management is increasingly relegated to lands less suitable for industrial agriculture or plantation forestry due to remoteness, nutrient impoverishment, steepness, or poor drainage (e.g. Putz et al., 2018). Remoteness generally increases the likelihood of governance failures while the adverse site conditions render forest lands more susceptible to soil damage and erosion. With the increased intensities of rainfall due to global climate change, soil compaction and erosion, including landslides, also increase, especially on steep slopes (Lele, 2009). The connections between these abuses in the hinterlands and downstream flooding need to be emphasized to spur improved enforcement of land-use regulations even in remote areas.

We recommend that to foster retention of the renewable natural resources and ecosystem services provided by tropical forests, the various values of forests should be disaggregated, considered individually, and then combined in an explicit manner to provide an overall evaluation of the sustainability of forest use at landscape scales. Increased transparency about the trade-offs associated with management decisions at stand up to landscape scales will at least inform debates. To increase the likelihood of political and behavioral changes that lead to improved fates of tropical forests, we advocate for the collaborative construction of detailed and place-specific theories-of-change in which the assumptions are enumerated, relevant actors are identified, their motivations and interactions are captured, and the contexts in which decisions are made are elucidated.

## 8  Ways forward

One major impediment to sustainable forest management at landscape scales is lack of appropriately trained foresters with the political wherewithal to have their voices heard. This deficiency increases as the number of forestry schools declines almost everywhere partially due to the demonization of tree cutting, despite the canonization of tree planting. Although people will always need wood and wood products, support for improved forest management by international organizations is likewise weak. While the abuses tropical forests suffer from timber mining operations are scrutinized by researchers, few and mostly naïve solutions are offered due to inattention to the relevant factors and constraints. Forest owners, be they governments or communities, also need to forgo some short-term profits so that the renewable natural resources in tropical forests have the chance to be renewed. Perhaps recognition that forest landscapes can be managed sustainably, without denying the many trade-offs, may help efforts to recruit motivated young people into the vibrant field of forestry.

## 9  References

Anitha, K., Joseph, S., Chandran, R. J., Ramasamy, E. V. and Prasad, S. N. 2010. Tree species diversity and community composition in a human-dominated tropical forest of western Ghats biodiversity hotspot India. *Ecological Complexity* 7(2), 217-24. doi:10.1016/j.ecocom.2010.02.005.

Baccini, A., Walker, W., Carvalho, L., Farina, M., Sulla-Menashe, D. and Houghton, R. A. 2017. Tropical forests are a net carbon source based on aboveground measurements of gain and loss. *Science* 358(6360), 230-4. doi:10.1126/science.aam5962.

Barlow, J., Gardner, T. A., Araujo, I. S., Avila-Pires, T. C., Bonaldo, A. B., Costa, J. E., Esposito, M. C., Ferreira, L. V., Hawes, J., Hernandez, M. I. M., Hoogmoed, M. S., Leite, R. N., Lo-Man-Hung, N. F., Malcolm, J. R., Martins, M. B., Mestre, L. A. M., Miranda-Santos, R., Nunes-Gutjahr, A. L., Overal, W. L., Parry, L., Peters, S. L., Ribeiro-Junior, M. A., da Silva, M. N. F., da Silva Motta, C. and Peres, C. A. 2007. Quantifying the biodiversity value of tropical primary, secondary, and plantation forests. *Proceedings of the National Academy of Sciences of the United States of America* 104(47), 18555-60. doi:10.1073/pnas.0703333104.

Barlow, A. C. D., Smith, J. L. D., Ahmad, I. U., Hossain, A. N. M., Rahman, M. and Howlader, A. 2011. Female tiger Panthera tigris home range size in the Bangladesh Sundarbans: the value of this mangrove ecosystem for the species' conservation. *Oryx* 45(1), 125-8. doi:10.1017/S0030605310001456.

Barnes, R. F. W., Barnes, K. L., Alers, M. P. T. and Blom, A. 1991. Man determines the distribution of elephants in the rain forests of northeastern Gabon. *African Journal of Ecology* 29(1), 54-63. doi:10.1111/j.1365-2028.1991.tb00820.x.

BirdLife International. 2016. Harpia harpyja. The IUCN RED List of Threatened Species 2016 e.T22695998A93537912.

Boscolo, M. 2000. *Strategies for Multiple Use Management of Tropical Forests: An Assessment of Alternative Options.* CID Working Paper Series.

Bowler, D. E., Buyung-Ali, L. M., Healey, J. R., Jones, J. P. G., Knight, T. M. and Pullin, A. S. 2012. Does community forest management provide global environmental benefits and improve local welfare? *Frontiers in Ecology and the Environment* 10(1), 29-36. doi:10.1890/110040.

Brockerhoff, E. G., Jactel, H., Parrotta, J. A., Quine, C. P. and Sayer, J. 2008. Biodiversity and planted forests–oxymoron or opportunity? *Biodiversity and Conservation* 17(5), 925-51. doi:10.1007/s10531-008-9380-x.

Brundtland, G. H. 1987. *Our Common Future.* Report of the World Commission on Environment and Development. Oxford University Press, Oxford.

Bruner, A. G., Gullison, R. E., Rice, R. E. and Da Fonseca, G. A. 2001. Effectiveness of parks in protecting tropical biodiversity. *Science* 291(5501), 125-8. doi:10.1126/science.291.5501.125.

Burivalova, Z., Şekercioğlu, C. H. and Koh, L. P. 2014. Thresholds of logging intensity to maintain tropical forest biodiversity. *Current Biology* 24(16), 1893-8. doi:10.1016/j.cub.2014.06.065.

Campos-Arceiz, A. and Blake, S. 2011. Mega-gardeners of the forest–the role of elephants in seed dispersal. *Acta Oecologica* 37(6), 542-53. doi:10.1016/j.actao.2011.01.014.

CPF Collaborative Partnership on Forests. 2010. *Measuring Forest Degradation.* Available at: http://www.fao.org/3/i1802e/i1802e00.pdf.

de Avila, A. L., Ruschel, A. R., de Carvalho, J. O. P., Mazzei, L., Silva, J. N. M., do Carmo Lopes, M. M., Araujo, M. M., Dormann, C. F. and Bauhus, J. 2015. Medium-term dynamics of tree species composition in response to silvicultural intervention intensities in a tropical rain forest. *Biological Conservation* 191, 577-86. doi:10.1016/j.biocon.2015.08.004.

Dudley, N. 2005. Best practices for industrial plantations. In: Mansouran, S., Vallauri, D. and Dudley, N. (Eds), *Forest Restoration in Landscapes: Beyond Planting Trees*. Springer Science, New York, pp. 379-97.

Edwards, D. P., Larsen, T. H., Docherty, T. D. S., Ansell, F. A., Hsu, W. W., Derhé, M. A., Hamer, K. C. and Wilcove, D. S. 2011. Degraded lands worth protecting: the biological importance of Southeast Asia's repeatedly logged forests. *Proceedings of the Royal Society B* 278(1702), 82-90. doi:10.1098/rspb.2010.1062.

Edwards, D. P., Gilroy, J. J., Woodcock, P., Edwards, F. A., Larsen, T. H., Andrews, D. J. R., Derhé, M. A., Docherty, T. D. S., Hsu, W. W., Mitchell, S. L., Ota, T., Williams, L. J., Laurance, W. F., Hamer, K. C. and Wilcove, D. S. 2014. Land-sharing versus land-sparing logging: reconciling timber extraction with biodiversity conservation. *Global Change Biology* 20(1), 183-91. doi:10.1111/gcb.12353.

Ellis, P. W., Gopalakrishna, T., Goodman, R. C., Roopsind, A., Griscom, B., Umunay, P. M., Zalman, J., Ellis, E., Mo, K., Gregoire, T. G. and Putz, F. E. 2019. Climate-effective reduced-impact logging (RIL-C) can halve selective logging carbon emissions in tropical forests. *Forest Ecology and Management* 438, 255-66.

FAO. 2018. *Terms and Definitions FRA 2020*. Forest Resources Assessment Working Paper 188. Rome, Italy.

Fisher, B., Edwards, D. P., Larsen, T. H., Ansell, F. A., Hsu, W. W., Roberts, C. S. and Wilcove, D. S. 2011. Cost-effective conservation: calculating biodiversity and logging trade-offs in Southeast Asia. *Conservation Letters* 4(6), 443-50. doi:10.1111/j.1755-263X.2011.00198.x.

Franca, F. M., Frazão, F. S., Koraski, V., Louzada, J. and Barlow, J. 2017. Identifying thresholds of logging intensity on dung beetle communities to improve the sustainable management of Amazonian tropical forests. *Biological Conservation* 216, 115-22. doi:10.1016/j.biocon.2017.10.014.

Griscom, B., Adams, J., Ellis, P., Houghton, R. A., Lomax, G., Miteva, D. A., Schlesinger, W. H., Shoch, D., Woodbury, P., Zganjar, C., Blackman, A., Campari, J., Conant, R. T., Delgado, C., Elias, P., Hamsik, M., Kiesecker, J., Landis, E., Polasky, S., Putz, F. E., Sanderman, J., Siikamäki, J., Silvius, M., Wollenberg, L. and Fargione, J. 2017. Natural pathways to climate mitigation. *Proceedings of the National Academy of Sciences (USA)* 114, 11645-50.

Griscom, B. W., Burivalova, Z., Ellis, P. W., Halperin, J., Marthinus, D., Runting, R., Ruslandi, B., Wahyudi, B. and Putz, F. E. 2019. Reduced-impact logging in Borneo to minimize carbon emissions while preserving sensitive habitats and maintaining timber yields. *Forest Ecology and Management* 438, 176-85.

Hodgson, J. A., Moilanen, A., Wintle, B. A., Thomas, C. D. 2011. Habitat area, quality and connectivity: striking the balance for efficient conservation. *Journal of Applied Ecology* 48(1), 148-52. doi:10.1111/j.1365-2664.2010.01919.x.

ITTO. 2015. *Voluntary Guidelines for the Sustainable Management of Natural Tropical Forests*. ITTO Policy Development Series No. 20, Yokohama, Japan.

ITTO. 2016. *Criteria and Indicators for the Sustainable Management of Tropical Forests*. ITTO Policy Development Series No. 21, Yokohama, Japan.

Kinnaird, M. F., Sanderson, E. W., O'Brien, T. G., Wibisono, H. T. and Woolmer, G. 2003. Deforestation trends in a tropical landscape and implications for endangered large mammals. *Conservation Biology* 17(1), 245-57. doi:10.1046/j.1523-1739.2003.02040.x.

Kohl, M., Neupane, P. R.and Lotfiomran, N. 2017. The impact of tree age on biomass growth and carbon accumulation capacity: a retrospective analysis using tree ring data of three tropical tree species grown in natural forests of Suriname. *PLoS ONE* 12(8), e0181187. doi:10.1371/journal.pone.0181187.

Kukkonen, M. and Hohnwald, S. 2009. Comparing floristic composition in treefall gaps of certified conventionally managed and natural forests of northern Honduras. *Annals of Forest Science* 66(8), 809. doi:10.1051/forest/2009070.

Laufer, J., Michalski, F. and Peres, C. A. 2013. Assessing sampling biases in logging impact studies in tropical forests. *Tropical Conservation Science* 6, 16-34.

Laurance, W. F., Useche, D. C., Rendeiro, J., Kalka, M., Bradshaw, C. J., Sloan, S. P., Laurance, S. G., Campbell, M., Abernethy, K., Alvarez, P. and Arroyo-Rodriguez, V. 2012. Averting biodiversity collapse in tropical forest protected areas. *Nature* 489, 290-4.

Lele, S. 2009. Watershed services of tropical forests: from hydrology to economic valuation to integrated analysis. *Current Opinion in Environmental Sustainability* 1(2), 148-55. doi:10.1016/j.cosust.2009.10.007.

Lewis, O. T. 2009. Biodiversity change and ecosystem function in tropical forests. *Basic and Applied Ecology* 10(2), 97-102. doi:10.1016/j.baae.2008.08.010.

Lewis, S. L., Edwards, D. P. and Galbraith, D. 2015. Increasing human dominance of tropical forests. *Science* 349(6250), 827-32. doi:10.1126/science.aaa9932.

Lindenmayer, D. B., Laurance, W. F. and Franklin, J. F. 2012. Global decline in large old trees. *Science* 338(6112), 1305-6. doi:10.1126/science.1231070.

Liu, C. L. C., Kuchma, O. and Krutovsky, K. V. 2018. Mixed-species versus monocultures in plantation forestry: development, benefits, ecosystem services and perspectives for the future. *Global Ecology and Conservation* 15, e00419. doi:10.1016/j.gecco.2018. e00419.

Loveridge, A. J., Searle, A. W., Murindagomo, F. and Macdonald, D. W. 2007. The impact of sport-hunting on the population dynamics of an African lion population in a protected area. *Biological Conservation* 134(4), 548-58. doi:10.1016/j.biocon.2006.09.010.

Luckert, M. and Williamson, T. 2005. Should sustained yield be part of sustainable forest management? *Canadian Journal of Forest Research* 35(2), 356-64. doi:10.1139/x04-172.

Luyssaert, S., Schulze, E. D., Börner, A., Knohl, A., Hessenmöller, D., Law, B. E., Ciais, P. and Grace, J. 2008. Old growth forests as global carbon sinks. *Nature* 455(7210), 213-5. doi:10.1038/nature07276.

Messier, C., Tittler, R., Kneeshaw, D. D., Gélinas, N., Paquette, A., Berninger, K., Rheault, H., Meek, P. and Beaulieu, N. 2009. TRIAD zoning in Quebec: experiences and results after 5 years. *The Forestry Chronicle* 85(6), 885-96. doi:10.5558/tfc85885-6.

Michalski, F. and Peres, C. A. 2013. Biodiversity depends on logging recovery time. *Science* 339(6127), 1521-2. doi:10.1126/science.339.6127.1521-b.

Mills, D. J., Bohlman, S. A., Putz, F. E. and Andreu, M. G. 2019. Liberation of future crop trees from lianas in Belize: completeness, costs, and timber-yield benefits. *Forest Ecology and Management* 439, 97-104. doi:10.1016/j.foreco.2019.02.023.

Mohebalian, P. M. and Aguilar, F. X. 2018. Beneath the canopy: tropical forests enrolled in conservation payments reveal evidence of less degradation. *Ecological Economics* 143, 64-73. doi:10.1016/j.ecolecon.2017.06.038.

Nasi, R. and Frost, P. G. H. 2009. Sustainable forest management in the tropics: is everything in order but the patient still dying? *Ecology and Society* 14(2), 40. Available at: www .ecologyandsociety.org/vol14/iss2/art40.

Nasi, R., Putz, F. E., Pacheco, P., Wunder, S. and Anta, S. 2011. Sustainable forest management and carbon in tropical Latin America: the case for REDD+. *Forests* 2(1), 200–17. doi:10.3390/f2010200.

Navarro-Martínez, A., Palmas-Perez, A. S., Ellis, E. A., Blanco Reyes, P., Vargas Godínez, C., Iuit Jiménez, A. C., Hernández Gómez, I., Ellis, P., Álvarez Ugalde, A., Carrera Quirino, Y. G., Armenta Montero, S. and Putz, F. E. 2017. Remnant trees in enrichment planted gaps Quintana Roo, Mexico: reasons for retention and effects on planted seedling growth. *Forests* 8, 272. doi:10.3390/f8080272.

Norden, N., Chazdon, R. L., Chao, A., Jiang, Y. H. and Vílchez-Alvarado, B. 2009. Resilience of tropical rain forests: tree community reassembly in secondary forests. *Ecology Letters* 12(5), 385–94. doi:10.1111/j.1461-0248.2009.01292.x.

Pan, Y., Birdsey, R. A., Fang, J., Houghton, R., Kauppi, P. E., Kurz, W. A., Phillips, O. L., Shvidenko, A., Lewis, S. L., Canadell, J. G., Ciais, P., Jackson, R. B., Pacala, S. W., McGuire, A. D., Piao, S., Rautiainen, A., Sitch, S. and Hayes, D. 2011. A large and persistent carbon sink in the world's forests. *Science* 333(6045), 988–93. doi:10.1126/science.1201609.

Paquette, A. and Messier, C. 2010. The role of plantations in managing the world's forests in the Anthropocene. *Frontiers in Ecology and the Environment* 8(1), 27–34. doi:10.1890/080116.

Payn, T., Carnus, J. M., Freer-Smith, P., Kimberley, M., Kollert, W., Liu, S., Orazio, C., Rodriguez, L., Silva, L. N. and Wingfield, M. J. 2015. Changes in planted forests and future global implications. *Forest Ecology and Management* 352, 57–67. doi:10.1016/j.foreco.2015.06.021.

Peña-Claros, M., Fredericksen, T. S., Alarcón, A., Blate, G. M., Choque, U., Leaño, C., Licona, J. C., Mostacedo, B., Pariona, W., Villegas, Z. and Putz, F. E. 2008. Beyond reduced-impact logging: silvicultural treatments to increase growth rates of tropical trees. *Forest Ecology and Management* 256(7), 1458–67. doi:10.1016/j.foreco.2007.11.013.

Phalan, B., Onial, M., Balmford, A. and Green, R. E. 2011. Reconciling food production and biodiversity conservation: land sharing and land sparing compared. *Science* 333(6047), 1289–91. doi:10.1126/science.1208742.

Pinard, M. A. and Putz, F. E. 1996. Retaining forest biomass by reducing logging damage. *Biotropica* 28(3), 278–95. doi:10.2307/2389193.

Piponiot, C., Rödig, E., Putz, F. E., Rutishauser, E., Sist, P., Ascarrunz, N., Blanc, L.,Derroire, G., Descroix, L., Laurent, C. G., Marcelino, H. C., Honorio Coronado, E., Huth, A., Kanashiro, M., Licona, J. C. and Mazzei, L. Neves d'Oliveira, M., Peña-Claros, M., Rodney, K., Shenkin, A., Rodrigues de Souza, C., Vidal, E., West, T., Wortel, V. and Hérault, B. 2019. Can timber provision from Amazonian production forests be sustainable? *Environmental Research Letters* 14(6), 064014. Available at: https://iopscience.iop.org/article/10.1088/1748-9326/ab195e.

Porter-Bolland, L., Ellis, E. A., Guariguata, M. R., Ruiz-Mallén, I., Negrete-Yankelevich, S. and Reyes-García, V. 2012. Community managed forests and forest protected areas: an assessment of their conservation effectiveness across the tropics. *Forest Ecology and Management* 268, 6–17. doi:10.1016/j.foreco.2011.05.034.

Potapov, P., Hansen, M. C., Laestadius, L., Turubanova, S., Yaroshenko, A., Thies, C., Smith, W., Zhuravleva, I., Komarova, A., Minnemeyer, S. and Esipova, E. 2017. The last frontiers of wilderness: tracking loss of intact forest landscapes from 2000 to 2013. *Science Advances* 3(1), e1600821. doi:10.1126/sciadv.1600821.

Puettmann, K. J., Wilson, S. M., Baker, S. C., Donoso, P. J., Droessler, L., Armente, G., Harvey, B. D., Knoke, T., Lu, Y., Nocentini, S., Putz, F. E., Yoshida, T. and Bauhus, J. 2015. Silvicultural alternatives to conventional even-aged management—what limits global adoption? *Forest Ecosystems* 2(1), 8. doi:10.1186/s40663-015-0031-x.

Putz, F. E. 2018. Sustainable = good, better, or responsible. *Journal of Tropical Forest Science* 30(1), 1–8. doi:10.26525/jtfs2018.30.1.18.

Putz, F. E. and Redford, K. H. 2010. Tropical forest definitions, degradation, phase shifts, and further transitions. *Biotropica* 42(1), 10–20. doi:10.1111/j.1744-7429.2009.00567.x.

Putz, F. E. and Romero, C. 2014. Futures of tropical forests (*sensu lato*). *Biotropica* 46(4), 495–505. doi:10.1111/btp.12124.

Putz, F. E., Dykstra, D. P. and Heinrich, R. 2000. Why poor logging practices persist in the tropics. *Conservation Biology* 14(4), 951–6. doi:10.1046/j.1523-1739.2000.99137.x.

Putz, F. E., Zuidema, P. A., Synnott, T., Peña-Claros, M., Pinard, M. A., Sheil, D., Vanclay, J. K., Sist, P., Gourlet-Fleury, S., Griscom, B., Palmer, J. and Zagt, R. 2012. Sustaining conservation values in selectively logged tropical forests: the attained and the attainable. *Conservation Letters* 5(4), 296–303. doi:10.1111/j.1755-263X.2012.00242.x.

Putz, F. E., Ruslandi, P., Ellis, P. W. and Griscom, B. 2018. Topographic restrictions on land-use practices: consequences of different pixel sizes and data sources for natural forest management in the tropics. *Forest Ecology and Management* 422, 108–13.

Putz, F. E., Baker, T., Griscom, B. W., Gopalakrishna, T., Roopsind, A., Umunay, P. M., Zalman, J., Ellis, E. A., Ellis, P. W. and Ellis, P. W. 2019. Intact forest in selective logging landscapes in the tropics. *Frontiers in Forests and Global Change* 2, 30. doi:10.3389/ffgc.2019.00030.

Rodrigues, A. S., Akcakaya, H. R., Andelman, S. J., Bakarr, M. I., Boitani, L., Brooks, T. M., Chanson, J. S., Fishpool, L. D., Da Fonseca, G. A., Gaston, K. J. and Hoffmann, M. 2004. Global gap analysis: priority regions for expanding the global protected-area network. *BioScience* 54, 1092–100.

Romero, C. and Putz, F. E. 2018. Theory-of-change development for evaluation of Forest Stewardship Council certification of sustained timber yields from natural forests in Indonesia. *Forests* 9(9), 547. doi:10.3390/f9090547.

Roopsind, A., Caughlin, T. T., Sambhu, H., Fragosa, J. M. V. and Putz, F. E. 2017. Logging and indigenous hunting impacts on persistence of large Neotropical animals. *Biotropica* 49(4), 565–75. doi:10.1111/btp.12446.

Roopsind, A., Caughlin, T. T., van der Hout, P., Arets, E. and Putz, F. E. 2018. Trade-offs between carbon stocks and timber recovery in tropical forests are mediated by logging intensity. *Global Change Biology* 24(7), 2862–74. doi:10.1111/gcb.14155.

Runting, R. K., Ruslandi, R., Griscom, B. W., Struebig, M. J., Satar, M., Meijaard, E., Burivalova, Z., Cheyne, S. M., Deere, N. J., Game, E. T., Putz, F. E., Wells, J. A., Wilting, A., Acrenaz, M., Ellis, P., Khan, F. A. A., Leavitt, S. M., Marshall, A. J., Possingham, H. P., Watson, J. E. M. and Venter, O. 2019. Larger gains from improved management over sparing-sharing for tropical forests. *Nature Sustainability* 2(1), 53–61. doi:10.1038/s41893-018-0203-0.

Ruslandi, W., Cropper, W. P. and Putz, F. E. 2017a. Effects of silvicultural intensification on timber yields, carbon dynamics, and tree species composition in a dipterocarp forest in Kalimantan, Indonesia: an individual-tree-based model simulation. *Forest Ecology and Management* 390, 104–18. doi:10.1016/j.foreco.2017.01.019.

Ruslandi, C., Romero, C. and Putz, F. E. 2017b. Financial viability and carbon payment potential of large-scale silvicultural intensification in logged dipterocarp forest in Indonesia. *Forest Policy and Economics* 85, 95–102. doi:10.1016/j.forpol.2017.09.005.

Sabogal, C., Guariguata, M. R., Broadhead, J., Lescuyer, G., Savilaakso, S., Essoungou, N. and Sist, P. 2013. *Multiple-Use Forest Management in the Humid Tropics: Opportunities and Challenges for Sustainable Forest Management.* FAO Forestry Paper No. 173. Food and Agriculture Organization of the United Nations, Rome, and Center for International Forestry Research, Bogor, Indonesia.

Santika, T., Meijaard, E., Budiharta, S., Law, E. A., Kusworo, A., Hutabarat, J. A., Indrawan, T. P., Struebig, M., Raharjo, S., Huda, I., Andini, S., Ekaputri, A. D., Trison, S., Stigner, M. and Wilson, K. A. 2017. Community forest management in Indonesia: avoided deforestation in the context of anthropogenic and climate complexities. *Global Environmental Change* 46, 60–71. doi:10.1016/j.gloenvcha.2017.08.002.

Sasaki, N. and Putz, F. E. 2009. Critical need for new definitions of "forest" and "forest degradation" in global climate change agreements. *Conservation Letters* 2(5), 226–32. doi:10.1111/j.1755-263X.2009.00067.x.

Sasaki, N., Asner, G. P., Pan, Y., Knorr, W., Durst, P. B., Ma, H. O., Abe, I., Lowe, A. J., Koh, L. P. and Putz, F. E. 2016. Sustainable management of tropical forests can reduce carbon emissions and stabilize timber production. *Frontiers in Environmental Science* 4. doi:10.3389/fenvs.2016.00050.

Sayer, J. A., Margules, C., Boedhihartono, A. K., Sunderland, T., Langston, J. D., Reed, J., Riggs, R., Buck, L. E., Campbell, B. M., Kusters, K., Elliott, C., Minang, P. A., Dale, A., Purnomo, H., Stevenson, J. R., Gunarso, P. and Purnomo, A. 2016. Measuring the effectiveness of landscape approaches to conservation and development. *Sustainability Science* 12(3), 465–76. doi:10.1007/s11625-016-0415-z.

Schaafsma, M., Burgess, N. D., Swetnam, R., Ngaga, Y., Ngowi, S., Turner, K. and Treue, T. 2013. Tanzanian timber markets provide early warnings of logging down the timber chain. In: *15th Annual BIOECON Conference, Conservation and Development: Exploring Conflicts and Challenges*, Cambridge, UK, pp. 18–20.

Schulze, M., Vidal, E., Grogan, J., Zweed, J. and Zarin, D. 2005. Madeiras nobres em perigo. *Revista Ciência Hoje* 36, 66–9.

Schulze, M., Grogan, J., Landis, R. M. and Vidal, E. 2008a. How rare is too rare to harvest? *Forest Ecology and Management* 256(7), 1443–57. doi:10.1016/j.foreco.2008.02.051.

Schulze, M., Grogan, J., Uhl, C., Lentini, M. and Vidal, E. 2008b. Evaluating ipê (*Tabebuia*, Bignoniaceae) logging in Amazonia: sustainable management or catalyst for forest degradation? *Biological Conservation* 141(8), 2071–85. doi:10.1016/j.biocon.2008.06.003.

Sebben, A. M., Degen, B., Azevedo, V. C. R., Silva, M. B., de Lacerda, A. E. B., Ciampi, A. Y., Kanashiro, M., Carneiro, F. S., Thompson, I. and Loveless, M. D. 2008. Modelling the long-term impacts of selective logging on genetic diversity and demographic structure of four tropical tree species in the Amazon forest. *Forest Ecology and Management* 254(2), 335–49. doi:10.1016/j.foreco.2007.08.009.

Sedjo, R. A. and Botkin, D. 1997. Using forest plantations to spare natural forests. Environment: Science and Policy for Sustainable Development 39: 14–30.

Slik, J. W. F., Paoli, G., McGuire, K., Amaral, I., Barroso, J., Bastian, M., Blanc, L., Bongers, F., Boundja, P., Clark, C., Collins, M., Dauby, G., Ding, Y., Doucet, J., Eler, E., Ferreira, L., Forshed, O., Fredriksson, G., Gillet, J., Harris, D., Leal, M., Laumonier, Y., Malhi, Y., Mansor, A., Martin, E., Miyamoto, K., Araujo-Murakami, A., Nagamasu, H., Nilus, R., Nurtjahya, E., Oliveira, Á, Onrizal, O., Parada-Gutierrez, A., Permana, A., Poorter, L., Poulsen, J., Ramirez-Angulo, H., Reitsma, J., Rovero, F., Rozak, A., Sheil,

D., Silva-Espejo, J., Silveira, M., Spironelo, W., ter Steege, H., Stevart, T., Navarro-Aguilar, G. E., Sunderland, T., Suzuki, E., Tang, J., Theilade, I., van der Heijden, G., van Valkenburg, J., Van Do, T., Vilanova, E., Vos, V., Wich, S., Wöll, H., Yoneda, T., Zang, R., Zhang, M. and Zweifel, N. 2013. Large trees drive forest aboveground biomass variation in moist lowland forests across the tropics. *Global Ecology and Biogeography* 22(12), 1261–71. doi:10.1111/geb.12092.

Sist, P., Mazzei, L., Blanc, L. and Rutishauser, E. 2014. Large trees as key elements of carbon storage and dynamics after selective logging in the Eastern Amazon. *Forest Ecology and Management* 318, 103–9. doi:10.1016/j.foreco.2014.01.005.

Solow, R. M. 1956. A contribution to the theory of economic growth. *The Quarterly Journal of Economics* 70(1), 65–94. doi:10.2307/1884513.

Sprecht, M. J., Pinto, S. R. P., Albuquerque, U. P., Tabarelli, M. and Melo, F. P. L. 2015. Burning biodiversity: fuelwood harvesting causes forest degradation in human-dominated tropical landscapes. *Global Ecology and Conservation* 3, 200–9. doi:10.1016/j.gecco.2014.12.002.

Stickler, C., Duchelle, A. E., Nepstad, D. and Ardila, J. P. 2018. Subnational jurisdictional approaches policy innovation and partnerships for change. In: Angelsen, A., Martius, C., De Sy, V., Duchelle, A. E., Larson, A. M. and Pham, T. T. (Eds), *Transforming REDD+: Lessons and New Directions*. CIFOR, Bogor, Indonesia.

Thompson, I. D., Mackey, B., McNulty, S. and Mosseler, A. 2009. *Forest Resilience, Biodiversity, and Climate Change*. A synthesis of the biodiversity/resilience/stability relationship in forest ecosystems. Secretariat of the Convention on Biological Diversity, Montreal. Technical Series no. 43, 67pp.

Thompson, I. D., Okabe, K., Parrotta, J. A., Brockerhoff, E., Jactel, H., Forrester, D. I. and Taki, H. 2014. Biodiversity and ecosystem services: lessons from nature to improve management of planted forests for REDD-plus. *Biodiversity and Conservation* 23(10), 2613–35. doi:10.1007/s10531-014-0736-0.

Thompson, I. D., Guariguata, M. R., Okabe, K., Bahamondez, C., Nasi, R., Heymell, V. and Sabogal, C. 2013. An operational framework for defining and monitoring forest degradation. *Ecology and Society* 18(2), 20. doi:10.5751/ES-05443-180220.

Van Gardingen, P. R., McLeish, M. J., Phililips, P. D., Fadilah, D., Tyrie, G. and Yasman, I. 2003. Financial and ecological analysis of management options for logged-over dipterocarp forests in Indonesian Borneo. *Forest Ecology and Management* 183(1–3), 1–29. doi:10.1016/S0378-1127(03)00097-5.

Vásquez-Grandón, A., Donoso, P. and Gerding, V. 2018. Forest degradation: when is a forest degraded? *Forests* 9(11), 726. doi:10.3390/f9110726.

Veldman, J. W., Overbeck, G. E., Negreiros, D., Mahy, G., Le Stradic, S., Fernandes, G. W., Durigan, G., Buisson, E., Putz, F. E. and Bond, W. J. 2015. Where tree planting and forest expansion are bad for biodiversity and ecosystem services. *BioScience* 65(10), 1011–8. doi:10.1093/biosci/biv118.

Vidal, E., West, T. A. P. and Putz, F. E. 2016. Recovery of biomass and merchantable timber volumes twenty years after conventional and reduced-impact logging in Amazonian Brazil. *Forest Ecology and Management* 376, 1–8. doi:10.1016/j.foreco.2016.06.003.

Vidal, E., West, T. A. P., Lentini, M. W., de Souza, S. E. X. F., Klauberg, C. and Waldhoff, P. 2020. *Sustainable Forest Management in the Brazilian Amazon*.

Vincent, J. R. and Binkley, C. S. 1993. Efficient multiple-use forestry may require land-use specialization. *Land Economics* 69(4), 370. doi:10.2307/3146454.

Watson, J. E. M., Evans, T., Venter, O., Williams, B., Tulloch, A., Stewart, C., Thompson, I., Ray, J. C., Murray, K., Salazar, A., McAlpine, C., Potapov, P., Walston, J., Robinson, J. G., Painter, M., Wilkie, D., Filardi, C., Laurance, W. F., Houghton, R. A., Maxwell, S., Grantham, H., Samper, C., Wang, S., Laestadius, L., Runting, R. K., Silva-Chávez, G. A., Ervin, J. and Lindenmayer, D. 2018. The exceptional value of intact forest ecosystems. *Nature Ecology and Evolution* 2(4), 599–610. doi:10.1038/s41559-018-0490-x.

Wiersum, K. F. 1995. 200 years of sustainability in forestry: lessons from history. *Environmental Management* 19(3), 321–9. doi:10.1007/BF02471975.

Wikramanayake, E., Dinerstein, E., Seidensticker, J., Lumpkin, S., Pandav, B., Shrestha, M., Mishra, H., Ballou, J., Johnsingh, A. J. T., Chestin, I., Sunarto, S., Thinley, P., Thapa, K., Jiang, G., Elagupillay, S., Kafley, H., Pradhan, N. M. B., Jigme, K., Teak, S., Cutter, P., Aziz, M. A. and Than, U. 2011. A landscape-based conservation strategy to double the wild tiger population. *Conservation Letters* 4(3), 219–27. doi:10.1111/j.1755-263X.2010.00162.x.

Wittemyer, G., Elsen, P., Bean, W. T., Burton, A. C. O. and Brashares, J. S. 2008. Accelerated human population growth at protected area edges. *Science* 321(5885), 123–6. doi:10.1126/science.1158900.

Zimmerman, B. L. and Kormos, C. 2012. Prospects for sustainable logging in tropical forests. *BioScience* 62, 479–87.

# Chapter 3

## Ecosystem services delivered by tropical forests: regulating services of tropical forests for climate and hydrological cycles

*Oliver Gardi, Bern University of Applied Sciences and School of Agricultural, Forest and Food Sciences HAFL, Switzerland*

1 Introduction

2 Forest-climate interactions

3 Forests in the carbon cycle

4 Climate change mitigation in the forestry and timber sector

5 Forests in the water cycle (regional scale)

6 Summary and future trends

7 Where to look for further information

8 References

## 1 Introduction

Forest ecosystem services, all the goods and services provided by forest ecosystems to the society, are manifold. They range from the tangible provision of timber and energy wood to biological diversity and further to intangible cultural and spiritual values. In the last couple of decades, the regulating services of forest ecosystems have gained increasing attention. In particular, the importance of forests for regulating the climate, hydrological cycles and maintaining soil fertility have become increasingly prominent in global environmental politics as well as in research.

In the context of global warming, particular focus is currently paid on the climate services provided by forests, mainly its role in the global carbon cycle. Forests exchange large quantities of carbon dioxide ($CO_2$) with the atmosphere and play a crucial role in the global carbon cycle. About one-tenth of the anthropogenic greenhouse gas emissions are caused by carbon dioxide released through deforestation of tropical forests and the associated depletion of terrestrial carbon stocks.

http://dx.doi.org/10.19103/AS.2020.0074.14

At the same time, forests currently also reabsorb about one-fourth of all anthropogenic carbon dioxide emissions, making forests globally an important net sink of $CO_2$. But, as forests themselves heavily depend on the climatic conditions, alterations in the climate system will itself affect forests' capacity to sequester carbon dioxide in the future.

Besides the direct exchange of $CO_2$ with the atmosphere through sinks and sources, forests also have indirect effects on the carbon cycle. Using wood for material, or as an energy source, reduces the demand for fossil fuels and for energy-intensive materials. This reduces $CO_2$ emissions by replacing fossil fuel use directly and in the manufacture of materials.

The role of forests in the global carbon cycle and the options for mitigating the greenhouse effect through forest management are complex. Forests are both sources and sinks of carbon dioxide; they provide wood (a key resource for the green economy) but all the interactions with the carbon cycle climate services are themselves, to a large extent, determined by climatic conditions.

To make things even more complex, the exchange of carbon dioxide with the atmosphere is not the only way that forests influence the climate system. Forest canopies are often darker than surrounding land-cover and thus absorb more energy from the solar radiation. Forests also have an effect through the quantity and type of forest exudates, which themselves influence the climate in different ways. Probably, most importantly, forests strongly influence the regulation of the global water cycle, with water vapour being the most important greenhouse gas, and precipitation/water availability the most limiting factor for terrestrial carbon uptake.

Although there was significant progress from research in the last couple of years on all these issues, we are still far from understanding all these processes and particularly their non-linear interactions. Such understanding is critical for a comprehensive appreciation of the role forests play in the global climate system and the ways in which forests can contribute to addressing global warming, probably the biggest challenge of our generation.

Nonetheless, enough is known on how to adjust forest management to address this challenge. Through preserving natural forests we can maintain forest carbon stocks, stimulate and stabilize hydrological cycles. Intensive but sustainable extraction of wood provides us with a renewable source of energy and material. With silvicultural measures we can assist the adaptation of forests for future climate conditions and, finally, we can, to a certain extent, increase the forest area in places suitable for reforestation.

Sound management of tropical forests is of particular importance for addressing the challenge of global warming for the following reasons: (a) the tropics harbour the largest forest areas and highest forest carbon stocks; (b) tropical rainforests are of tremendous importance for regional hydrological regimes; (c) they are extremely vulnerable to changes in climate and water regimes and could react by sudden dieback as soon as tipping points are reached; and (d) they face high pressure from commercial agriculture as well as from meeting the food and energy demands of a rapidly increasing population.

Global warming is significantly altering the environmental conditions in which forests grow at a global scale. At the local scale, the impact of global warming on forests depends heavily on current forest type and structure, soil conditions and water availability (Fig. 1). Managing forests for the provision of ecosystem services requires anticipating how global warming might alter local environmental conditions and how this will affect forests. Based on this anticipation, measures can be taken to reduce the risks and support forests in their adaption to those anticipated conditions. Such measures should take into account uncertainties and the effect that those measures themselves will have on climatic and other environmental conditions.

This chapter aims to contribute to this process by summarizing the current state of knowledge on the interactions between forest ecosystems and the climate system and the way in which forests influence the water cycle.

## 2  Forest-climate interactions

Forests influence the climate and climate influences forests. As forest-climate interactions are bidirectional, climatic changes will have consequences for forests that then influence the climate. In general, there are two types of retroactions: negative feedback loops, which regulate a system and contribute to its stabilization, and positive feedback loops, which reinforce the cause of change. The forest-climate interaction is considered to be a negative feedback loop, helping to naturally regulate the global climate system.

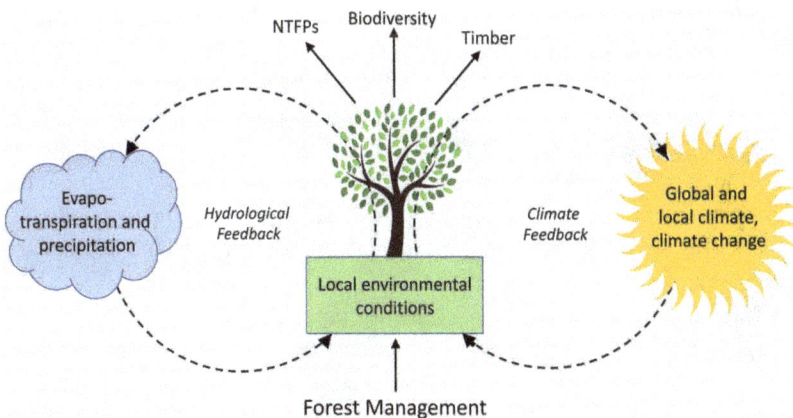

**Figure 1** Managing forests for maintaining a continuous flow of ecosystem services needs to take into account the complex interactions of forests with the environment.

On a geological time-scale, forests played an important role as part of earth's thermostat. Whenever the earth was warming, forest cover increased, carbon dioxide was sequestered from the atmosphere and the greenhouse effect was reduced. On the other hand, when the earth was cooling, forest area decreased, large quantities of carbon dioxide were released and the greenhouse effect was stimulated.

Applying these general relations observed in the earth's history to current global climate change would be wrong for two reasons. First, global temperature is currently warming at unprecedented rates, surpassing by far the adaptive capacity of forest ecosystems and could thus potentially have negative rather than positive effects on forest growth. Secondly, forest-cover, as is any other land-cover, often far from natural and strongly influenced by anthropogenic activities while forests' ability to expand and migrate, is very limited. In order to manage forests for human well-being in general and specifically for mitigating climate change, it is thus important to look in more detail at forest-climate interactions and how these might be affected by climate change and changes in land-use.

There is no single interaction between forests and the climate system. Instead, forests and climate interact in complex ways through multiple biophysical and biogeochemical feedback, each of these resulting in warming or cooling across different spatial and temporal scales. Biogeochemical forest-climate interactions relate mainly to the alteration of the atmospheric concentrations of carbon dioxide due to changes in forest carbon stocks.

Growing forests sequester carbon dioxide from the atmosphere and thereby contribute to cooling due to a reduction of the greenhouse effect. Releasing the carbon stored in the biomass of world's forest would increase global temperatures by about 1°C. However, an increase in the forest area increases atmospheric water vapour, another greenhouse gas. Although this effect is not yet fully understood, it is thought that the associated heating from this partly compensates for the cooling effect provided by carbon sequestration.

The biogeochemical cooling that growing forests provide through absorption of carbon dioxide is modulated by biophysical forest-climate interactions that influence the energy transfer and that can have both cooling and warming effects. The main biophysical interactions comprise the influence of forests on cloud formation, surface reflectance (albedo) and the exchange of sensible and latent heat.

Through their influence on cloud formation, due to transpiration and emittance of forest exudates, forests influence the amount of solar irradiation and thermal emittance. Clouds enhance the reflection of solar radiation (albedo) and thus provide a cooling effect. On the other hand, however, clouds prevent thermal radiation loss and thereby contribute to warming. Evidence that clouds have a net positive feedback and thus reinforce the warming effect is increasing. However, in considering the effect on forests, different types of clouds have to be distinguished.

Dense stratus and cumulus clouds are often found above forests. These have particularly high reflectance, up to 90%, and thus provide a strong cooling effect, particularly in tropical regions where solar irradiation is higher than in higher latitudes. Furthermore, as these cloud types are located in low and warm atmospheric layers, their thermal emittance is similar to the thermal emittance of the earth surface and only have a marginal warming effect on surface temperature. Overall, they have a net cooling effect.

By contrast, the cooling effect of cirrus clouds due to the reflection of solar radiation is usually below 60% because of their semi-transparency. As cirrus clouds are located in higher and cooler atmospheric layers, their thermal radiation is lower than that of the earth's surface. Cirrus clouds, therefore, contribute to a warming of lower atmospheric layers and the earth's surface.

Besides their influence on the formation of clouds, forests also directly influence the reflectance of solar irradiation as their canopies are usually darker than the surrounding environment. While bare land has an albedo of about 30%, forests reflect only about 10-20% of incoming solar radiation, an effect known as the biophysical warming effect of forests. The difference of the albedo between forests and the non-forest environment is particularly pronounced in boreal regions, where in the winter, the dark coniferous forests are in stark contrast to the bright snow covering other land-cover types. In tropical regions, this effect is less pronounced.

Another biophysical forest-climate interaction relates to the exchange of heat between the earth's surface and the atmosphere. First, due to the roughness of the canopy surface, forests contribute to the mixing of lower atmospheric layers and thereby the transport of heat from the earth's surface to the atmosphere. Secondly, forests transpire large amounts of water, thereby absorbing the heat needed for evaporation at the earth's surface, which is released back into the atmosphere when the water vapour condenses.

Taking all this together, what is the effect of forests on global earth surface temperatures? While it is undisputed that forest growth contributes to a biogeochemical cooling at the global scale, due to their absorbing and storing carbon dioxide, the net global effect of the biophysical interactions is much more uncertain. It ranges from slight cooling to slight warming.

The Intergovernmental Panel on Climate Change estimates that emissions of carbon dioxide from a complete loss of tropical forests would cause a significant biogeochemical warming of $+0.53 \pm 0.32°C$, while the net warming effect of biophysical interactions is much lower, at $+0.1 \pm 0.27°C$, and not significant (IPCC SRCCL, Jia et al., 2019). The climate models found that in a deforestation scenario, the biophysical warming due to a decreased flux of sensible and latent heat is almost completely compensated by cooling effects due to a reduced albedo and decreased atmospheric water vapour.

While the biophysical forest-climate interactions might not have a significant effect on average global earth surfaces temperatures, they are of importance for the regional expression of global warming. This is because different biophysical effects are expressed on different spatial scales. While the cooling effect of deforestation due to the reduction of atmospheric water vapour is expressed over a large scale (similar to greenhouse gases), the warming effects of reduced heat exchange are mainly expressed locally. A scenario of tropical deforestation is thus expected to cause a significant regional warming of +0.61 ± 0.48°C, in addition to the biogeochemical global warming (IPCC SRCCL, Jia et al., 2019).

## 3    Forests in the carbon cycle

According to the current state of knowledge, the biogeochemical exchange of carbon dioxide with the atmosphere is the most important forest-climate interaction, particularly in the tropics, where biomass density of forests is generally high and the albedo difference with other land-cover is relatively low.

The process of sequestering carbon dioxide from the atmosphere through photosynthesis is called gross primary production (GPP). About half of the sequestered carbon is used for the energy supply of the plant itself through oxidation and release of carbon dioxide again (autotrophic respiration, $R_a$). The result of GPP minus autotrophic respiration is called net primary production (NPP). This is used for building up biomass in plant tissues. These tissues then provide food for animals and micro-organisms (heterotrophic respiration, $R_h$).

The net ecosystem production (NEP) is the sum of all these biological processes (Lovett et al, 2006). A broader definition of the NEP, sometimes referred to as net biome production (NBP), also includes the non-biological oxidation of carbon (mainly fires) and the physical import and export of organic material (e.g. through harvesting). The NBP is a comprehensive measure for net carbon accumulation by an ecosystem. It indicates whether the ecosystem as a whole is acting as a net sink or a net source of carbon dioxide, taking into account the carbon pools of living and dead biomass as well as soil organic matter:

$$\Delta C_{org} = GPP - R_a - R_h - Ox_{nb} + I - E$$

With about 850 GtC, global forests store about the same amount of carbon in biomass and soil as is contained as $CO_2$ in the atmosphere. More than half of this is stored in tropical forests (Pan et al., 2011). In general, the main forest carbon pools are soils (50%), above-ground biomass (30%), below ground biomass (10%), deadwood and litter (10%). In tropical forest, however, the proportion of carbon stored in biomass is significantly higher than the biomass stored in the soil (see Table 1).

Through photosynthesis and respiration, terrestrial ecosystems annually turn over about one-sixth of the atmosphere's carbon dioxide (see Fig. 2). Pre-industrial GPP of terrestrial ecosystems is estimated at 109 GtC/year while total respiration including fires is estimated at 107 GtC/year. The remaining NPP of 2 GtC/year remains in the system, and is finally exported as soil organic material through rivers. All terrestrial ecosystems together are in balance, neither acting as a net sink, nor as a net source of carbon dioxide.

Human activities altered the natural carbon cycle in general and the carbon balance of terrestrial ecosystems in particular. Land-cover change and land management have a direct impact on terrestrial carbon stocks. Indirect effects

**Table 1** Approximate areas of different forest types, with typical values of soil and above-ground biomass carbon densities, including annual net growth of above-ground carbon stocks. Area estimates are from FAO, 2001. AGB values are compiled from Table 4.12 of the 2019 Revision of IPCC AFOLU Guidelines (Domke et al., 2019). Multiply with two to obtain biomass values (t d.m. ha$^{-1}$) or with 44/12 for obtaining $CO_2$ equivalents (t$CO_2$ ha$^{-1}$). Soil carbon stocks are from Houghton and Nassikas (2017)

| | Area Mha | Soils tC ha$^{-1}$ | Primary forest AGB (tC ha$^{-1}$) | | Secondary forest AGB (tC ha$^{-1}$) | | Plantations AGB (tC ha$^{-1}$) | |
|---|---|---|---|---|---|---|---|---|
| **Tropical forests** | **1880** | | | +0 | | +1 | | +5 |
| • Rainforest | 1120 | 120 | 187 | | 60 | | 96 | |
| • Mountain systems | 160 | 75 | 136 | | 64 | | 44 | |
| • Moist deciduous forest | 440 | 100 | 82 | | 39 | | 82 | |
| • Dry forest | 200 | 40 | 64 | | 55 | | 36 | |
| • Shrublands | – | 35 | 26 | | 26 | | 25 | |
| **Subtropical forests** | **360** | | | +1 | | +1 | | +6 |
| • Humid forests | 160 | 120 | 77 | | 66 | | 70 | |
| • Mountain systems | 120 | 120 | 60 | | 44 | | 40 | |
| • Dry forests | 40 | 80 | 42 | | 42 | | 36 | |
| • Steppe | – | 50 | 23 | | 23 | | 19 | |
| **Temperate forests** | **440** | 190 | 148 | +1 | 62 | +2 | 32 | +2 |
| **Boreal forests** | **1320** | 206 | 31 | +1 | 40 | +1 | 18 | +1 |

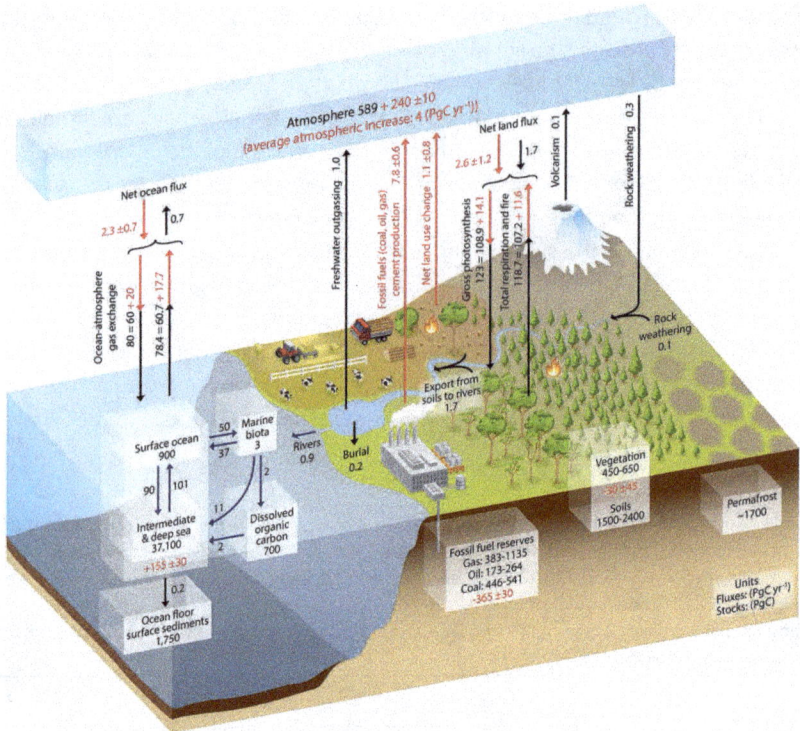

**Figure 2** The global carbon cycle and its anthropogenic perturbations. Black numbers and arrows show the pre-industrial stocks and annual flows. Red arrows and figures show anthropogenic modifications of the stocks since about 1750 and flows from 2000 to 2009 (source: IPCC AR5, Ciais et al., 2013, Figs 6-1).

of anthropogenic activities are changes in net ecosystem productivity due to altered environmental conditions.

Based on data from FAOSTAT and bookkeeping models, direct anthropogenic loss of terrestrial carbon stocks due to net land-use change between 2000 and 2010 is estimated at 1.1 GtC/year. Associated emissions correspond to 4 GtCO$_2$/year or about 12% of current anthropogenic CO$_2$ emissions (Houghton and Nassikas, 2017; Pan et al., 2011). This figure is dominated by net carbon dioxide emissions from deforestation in the tropics of 5 Gt CO$_2$/year, which is partly compensated by forest regrowth in temperate and boreal regions of 1 Gt CO$_2$/year. Further, it is estimated that intact remaining forests sequester about 9 GtCO$_2$/year, half of it by tropical forests. Taking this together, tropical forests are assumed to be a net source of CO$_2$ of about 0.5 GtCO$_2$/year.

Direct assessments, using satellite imagery, of changes in above-ground carbon storage result in somewhat lower estimates of gross emissions from deforestation in the tropics (without regrowth) of about 3 GtCO$_2$/year for the period 2000-2010 (Baccini et al., 2017; Harris et al., 2012). This corresponds to an annual loss of about 40 million ha of forest with an average biomass density of 75 tC/ha (see typical above-ground biomass carbon densities of tropical forests in Table 1). Only about half of these losses of terrestrial carbon stocks are compensated for by re-establishment of forests/plantations (0.3 GtCO$_2$/year) and growth in remaining forests (1.2 GtCO$_2$/year). Using this direct assessment, tropical forests are found to be a net source of CO$_2$ of about 1.5 GtCO$_2$/year.

While deforestation is one of the most important sources of anthropogenic CO$_2$ emissions, altered climatic conditions and other indirect anthropogenic effects are thought to have caused considerable gains in terrestrial carbon stocks over the last decade. On 25-50% of the earth's vegetated area, a 'greening' has been observed in the period 1982-2009, while a 'browning' was observed on only 4% of the area (Zhu et al., 2016). Contrary to previous views, Song et al. (2018) found that, globally, the area covered by trees increased between 1982 and 2016 and that losses of tree cover in the tropics are more than compensated by gains in the subtropical, temperate and boreal climate domains.

Dynamic global vegetation models (DGVMs) indicate that as a result of altered climate conditions (the prolonged vegetation period in higher latitudes) plus fertilization due to increased atmospheric carbon dioxide concentration and nitrogen deposition, the GPP of terrestrial ecosystems increased by about 13% (52 GtCO$_2$/year). Total respiration, including fires, increased only by 11% (42 GtCO$_2$/year). The result is an anthropogenically induced net sink of carbon dioxide of about 10 GtCO$_2$/year or about a quarter of total anthropogenic CO$_2$ emissions (Keenan et al., 2016). This residual sink is mainly attributed to old-growth tropical, temporal and boreal forests (in this order), and is two to three times higher than emissions caused by land-use change.

While global forests provide a large net CO$_2$ sink, it is not yet clear to what extent tropical forests contribute to this sink. Dynamic vegetation models (DVMs) expect tropical forests to now be sequestering larger amounts of CO$_2$ due to increased CO$_2$ concentrations in the atmosphere, thus compensating the losses from land-use change (Fig. 3). Alternatively, recent studies combining forest inventories with satellite imagery suggest that carbon stock increases in tropical forest are not sufficient to cover carbon stock losses and that tropical forests are a slightly net source of CO$_2$ (Baccini et al., 2017). If correct, this would mean that the current terrestrial land sink is mainly provided by temperate and boreal forests.

How future climate change will affect the productivity of forests is highly uncertain. While DGVMs agree on the increased productivity and net carbon dioxide sink observed in the past, they start to diverge when it comes to projections for the second half of the twenty-first century. As shown in Fig. 4,

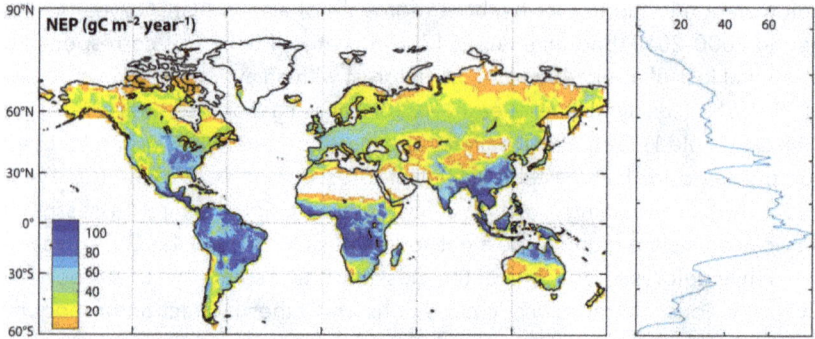

**Figure 3** Global distribution of annual net ecosystem production (NEP) (2000-2010) as modelled by ensembles of Dynamic Vegetation Models. (TRENDY model intercomparison project. Source: Keenan and Williams, 2018).

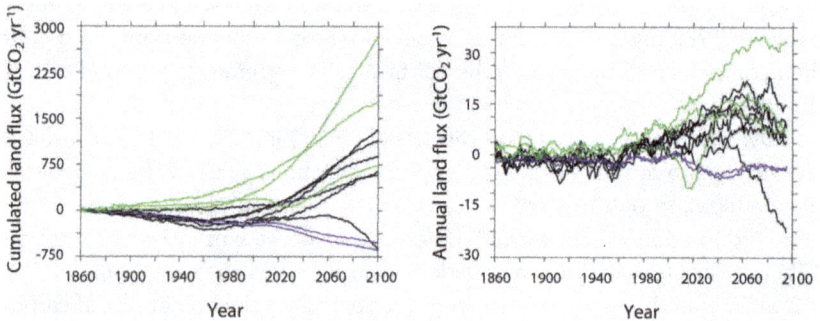

**Figure 4** Range of cumulative and annual atmosphere to land carbon fluxes as simulated by 11 Earth System Models in the CMIP5. Future fluxes are based on IPCC business-as-usual scenario RCP8.5 (Source: Friedlingstein, 2014).

most models predict gradual changes, with a peak in the forest sink around 2060 at about 14 $GtCO_2$/year that decreases thereafter to a net sink of 7 $GtCO_2$ by the end of the century, with a range between 33 $GtCO_2$/year and −22 $GtCO_2$/year (Friedlingstein et al., 2014).

It is known from observations in the past that forests react gradually to environmental changes, the nature of change as well as its dimension and timing. When certain thresholds are reached, the mainly hydrologically determined self-maintenance of forests cannot be maintained and forests might rapidly convert to steppe. Such sudden changes have been observed in the Sahara that has switched several times from forest to steppe and back again during the ice ages and even during the Holocene.

Rapid changes of forest cover indicate critical transitions between two stable states of the system (bi-stability, see Fig. 5). Once a perturbation causes the system to switch into the other stable state, it will not be possible to return

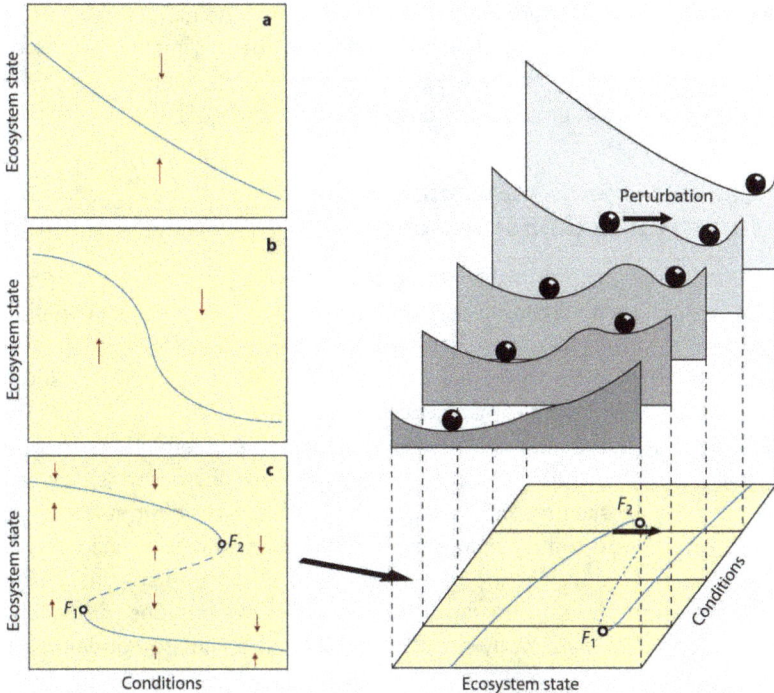

**Figure 5** Schematic representation of different system responses to gradual changes. The left panel shows different system types – (a) linear system: gradual, unsurprising and reversible response, (b) non-linear system: disproportionally strong responses, but still continuous and reversible and (c) critical or bi-stable system: abrupt, non-continuous and non-reversible responses. The right panel shows how in bi-stable systems perturbations of conditions might have no effect until the system reaches a tipping point F1 or F2. (modified from Scheffer et al. 2001).

to the initial state. Coupled atmospheric and vegetation models suggest that large parts of the Amazon are currently in a bi-stable state. When initialized with forest, the model developed towards rainforest; when initialized with desert, it developed towards savanna (Oyama and Nobre, 2003).

Transitions in bi-stable systems are usually of large dimension, but difficult to predict and thus to manage. Named as 'tipping points' they are also among the great unknowns in the climate system. Potential tipping points that involve forests are rapid dieback of boreal forests and Amazon rainforest dieback as well as greening of the Sahara and Sahel region due to shift in West African Monsoon (Lenton, 2013; Claussen, 2008).

The forest-climate interaction has traditionally been thought to have a stabilizing, negative feedback loop on atmospheric warming. As higher temperatures and increased $CO_2$ concentrations will support higher forest growth rates, this will tend to have a cooling effect through a reduced $CO_2$ concentration in the atmosphere. Depending on whether tipping points are reached, the outcome is uncertain, and rapid dieback of forests due to climate change would result in strong positive climate feedback. The enormous quantities of $CO_2$ this would release in short time would contribute to rapid global warming.

## 4    Climate change mitigation in the forestry and timber sector

Considering the role of forests in the global carbon cycle, the most important aspect for the forest sector at the global scale in climate change mitigation is to maintain the forest $CO_2$ sink or at least hinder its reversal. Remaining natural forests have to be maintained and measures have to be assured that support forest adaptation to future climatic conditions.

General recommendations for adaptive forest management are to reduce the vulnerability of forests and the damage from increased stress while increasing their resilience, the system's ability to recover after stress (Fig. 6). Both can be supported by increasing the structural diversity of forests, the species diversity, as well as the genetic diversity of the tree species. This can be done through the introduction of more resilient genotypes (provenances or improved varieties) already pre-adapted to future climate conditions. The

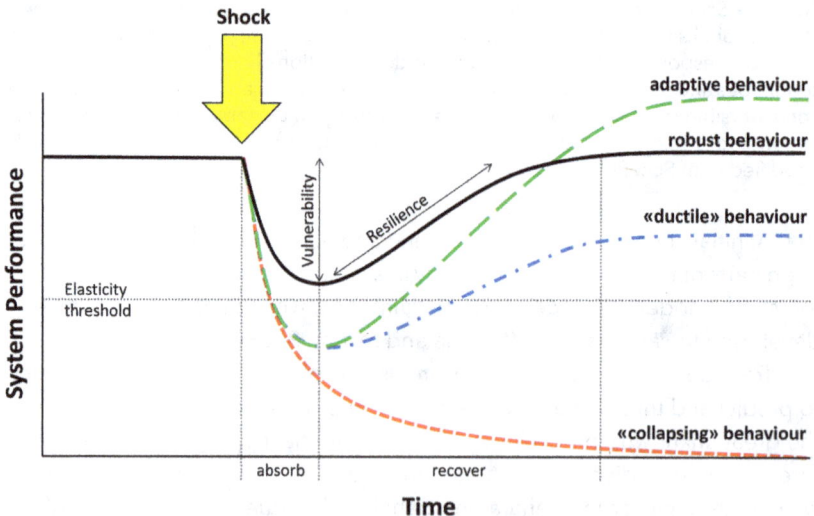

**Figure 6** Schematic illustration of system vulnerability and resilience. Vulnerability is the extent to which the system performance is reduced due to the stress, resilience the system's capacity to recover from stress. Up to a certain threshold, systems can resist to external stress and recover after a certain period. If the threshold is exceeded in intensity and/or duration, a system can react in different ways from collapse to even improved performance if the system is able to adapt to the new conditions.

economic risks of damage in managed forests can further be reduced through a reduction of the rotation period. A shorter rotation period further allows for regular interventions for increasing the diversity and reducing the risk of loss and damages (Brang et al., 2014).

In most tropical countries where forest areas are decreasing and the quality of forests is deteriorating due to over-use, the priority is not yet on adapting forests to climate change and improving forest productivity in the long run. It is primarily on maintaining the forest area and quality (reducing deforestation and degradation of forests) in the short term. Deforestation and degradation currently account for about 12% of global anthropogenic $CO_2$ emissions. The second priority is on increasing the forest area on degraded lands through forest landscape restoration and the regeneration and improvement of degraded forests.

Reducing deforestation and forest degradation by maintaining carbon pools and controlling disturbances such as fires and pests is seen as the most immediate and the less cost-intensive climate change mitigation option for the forest sector. Some 40 million ha of forest are converted to non-forest land uses annually, causing emissions of 3–5 $GtCO_2$/year. The potential for reducing emissions is estimated between 0.5 and 5.8 $GtCO_2$/year (IPCC SRCCL, Jia et al., 2019). In the long run, these restored and/or saved forests will further sequester $CO_2$ from the atmosphere and might be used for delivering wood for a green economy.

Practice, however, shows that maintaining forest areas is itself an enormous social challenge for many tropical countries. Poor people, in particular, depend on clearing forests for their livelihoods. From an economic perspective, it must be recognized that forests usually compete for land with alternative uses and the main underlying driver for forest conversion is the agricultural rent that can be obtained on the same parcel of land. More than 80% of the deforestation observed in the tropics can be attributed to agriculture, with the share of commercial agriculture rapidly increasing. From a governance perspective, reducing deforestation requires a balanced set of measures, including tenure reform, cross-sectoral collaboration, fair and transparent incentive systems and effective law enforcement.

Another option for increasing the $CO_2$ sink of forests is through afforestation/reforestation and regeneration of degraded forests. Globally, more than 2 billion ha of degraded land could be potentially suitable for forest restoration (Potapov et al., 2011). Societal challenges for implementing such activities might be lower, as most of this land has low productivity and is only marginally useful for livelihoods. However, initial investment would be high and $CO_2$ sequestration needs time. In total, however, due to the enormous area available, the potential for sequestering additional $CO_2$ is estimated at

0.5-10.1 $GtCO_2$/year, even higher than the potential for reducing emissions from deforestation and degradation (IPCC SRCCL, Jia et al., 2019).

A third and final mitigation option is in the forest itself (production side). Once forest area is secured and degraded land is restored, management of existing forests can improve productivity, turnover rates, harvest rates, carbon storage in wood products and soil carbon. The potential for reducing emissions or enhancing removals with improved forest management is estimated at 0.4–2.1 $GtCO_2$/year.

The climate benefit from forest management is closely linked with the climate benefit of using wood as a raw material and a source of energy, and trade-offs exist between these two. Timber and wood-based biomass play a central role in a carbon neutral economy by providing wood to replace energy-intensive materials and biomass to replace oil as raw material in petrochemical processes.

At the same time, managed forest can deliver net sequestering of $CO_2$ from the atmosphere by increasing carbon stocks in the forest and in long-lived wood products. There are two ways in which sustainably sourced wood can be used in order to reduce the use of fossil fuels while net sequestering $CO_2$ from the atmosphere at the same time. These are the increased use of wood for long-lived products and for bio-energy (or more general bio-chemistry) and the capture and long-term storage of $CO_2$ in the forests providing these.

Continuous removal of wood might lower the forest carbon stock compared with untouched forests, but it can help to maintain a healthy and productive forest. As long as the harvested wood is not burnt, the $CO_2$ remains stored in the wood products. Harvested wood products can thus be

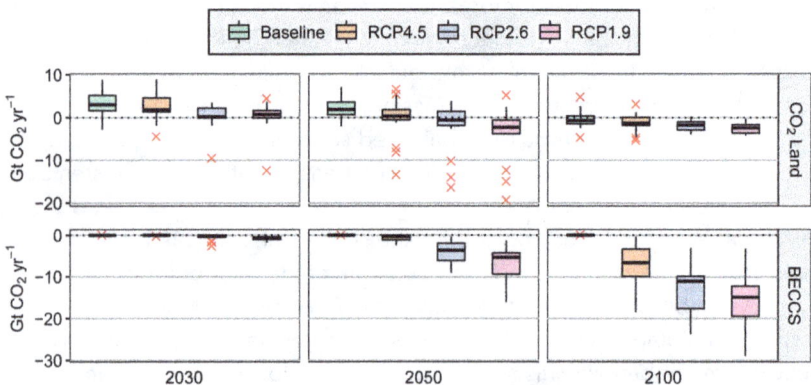

Figure 7 Net sequestration by land and bioenergy carbon capture and storage as assumed in the IPCC scenarios RCP8.2 (baseline), RCP4.5, RCP2.6 and RCP1.9 for the years 2030, 2050 and 2100 (Source: IPCC SRCCL, Jia et al., 2019).

considered as an extended forest carbon pool. At the end of life, the $CO_2$ from wood combustion could be captured and stored in the soil. Taken together, a sustainably managed forest combined with bio-energy with carbon capture and storage (BECCS) technology would provide a net sink of $CO_2$ from the atmosphere ('negative emissions').

Any scenario envisioning a stringent climate change mitigation target relies heavily on BECCS technology (Fig. 7). Scenarios expect that this technology will provide additional sequestration (or negative emissions) of between 6.5 and 15 $GtCO_2$/year. Assuming a net sequestration of about 3 $tCO_2$ $ha^{-1}$ $yr^{-1}$ around 500 million ha of land would be required.

Such high $CO_2$ removal through mitigation options based on land conversion (afforestation and BECCS) would shape the land system dramatically and would require an increase in the forest area of about 1 billion ha. This would be taken from pasture land and low production/degraded cropland. It is estimated that globally some 1.5 billion ha of potential forest land exists that could store about 750 $GtCO_2$ in areas that were previously degraded, dominated by sparse vegetation, grasslands and degraded bare soils (Bastin et al., 2019). However, large-scale reforestation requires substantial investment, good practices and it takes time for the full carbon removal to be achieved as the forest grows.

Using wood and timber as renewable raw materials and thereby replacing fossil fuel-based or energy-intensive materials such as cement or steel could reduce emissions of fossil $CO_2$ from the industrial and construction sectors. It is estimated that the use of 1 $m^3$ wood for material purposes on average avoids the emission of 0.5-1 $tCO_2$ from the production of alternative materials (1.2 kC/kgC, Leskinen et al., 2018).

With cascade use, where the same 1 $m^3$ of wood is recycled and used for multiple purposes (solid wood products -> particle-based products -> fibre-based products -> chemical products), this material substitution effect is replicated and increased with each utilization step. Finally, using this 1 $m^3$ of wood to reduce dependency on fossil fuels for energy generation avoids the emission of about 0.5-1 $tCO_2$ in the energy sector.

Example: Without harvesting, a forest with an annual increment of 5 $m^3$ $ha^{-1}$ $yr^{-1}$ would increase its forest carbon stock by up to 5 $tCO_2$ $ha^{-1}$ $yr^{-1}$ in the short run. In the long run, however, the increment would be balanced by increased mortality until increment and morality were in balance. Through sustainable harvesting, the extraction of 5 $m^3$ $ha^{-1}$ $yr^{-1}$ has the potential to provide a long-term net carbon sink of about 5 $tCO_2$ $yr^{-1}$ and to reduce fossil $CO_2$ emissions in other sectors by 5-10 $tCO_2$ $yr^{-1}$.

In a conclusion, two principal strategies exist with regard to carbon management in the forest and timber sector: Focussing on the preservation and enhancement of forest carbon stocks versus maximizing yields and efficient use of wood. Maintaining and increasing forest carbon stocks has immediate climate benefits. However, the forest carbon stock will reach a maximum at some point in the future and become vulnerable to reversal under future climate conditions.

Maximizing forest productivity and complete harvesting of the yield allows for maximizing the extraction of wood that can be (a) used for long-lived products such as timber, replacing energy-intensive materials such as concrete and steel; (b) for bioenergy substituting fossil fuels; (c) to be buried as biochar; or (d) to be used in the wider bio-economy. While this strategy might require the reduction of the initial forest carbon stocks, it enables forest land to be continuously used for mitigation in the long run (Fig. 8).

In general, we can say that, 'a sustainable forest management strategy aimed at maintaining or increasing forest carbon stocks, while producing an annual sustained yield of timber, fibre or energy from the forest, will generate the largest sustained mitigation benefit' (IPCC AR4, Nabuurs et al., 2007). The trade-off between maximizing forest carbon stocks and maximizing wood harvest for substitution, however, depends on the initial forest condition, growth rates, energy used for extraction and processing and on substitution efficiency (cascade use of wood).

So, while for degraded but highly productive and accessible forest lands close to the markets a management option maximizing the yield (e.g. *Eucalyptus* coppicing for wood energy) might be the best management option for mitigation, conservation might be the more optimal solution for a remote primary forest with very high carbon stocks, as the initial carbon stock losses and associated pay-back time through management would be too high.

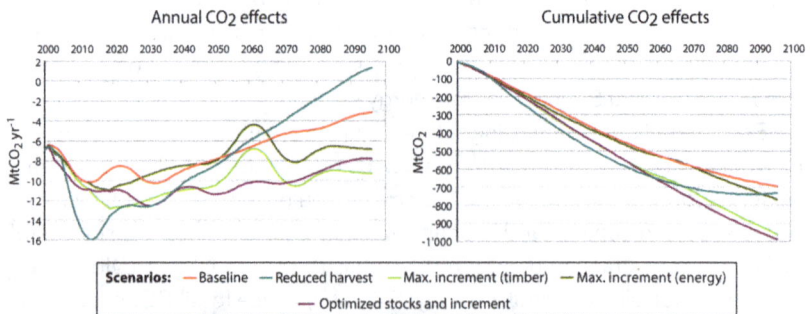

**Figure 8** Whole-economy $CO_2$ effects of different forest management options in Switzerland. Enhancement of carbon stocks in forests has the highest short-term effect. On the long run, however, (about 50 years in the above example) management strategies sustainably exploit timber for use in the building sector (source: Taverna et al., 2007).

# 5 Forests in the water cycle (regional scale)

Forests are not only influencing the climate system, but they are also heavily influenced by it. It is mainly the availability of water that determines where forest can grow. Plants need water in order to assimilate $CO_2$ and nutrients and trees usually need more water than other plants due to their height, leaf area and roughness – the growth of closed forests requires an annual rainfall of about 500 mm in boreal regions and about 1500 mm in the tropics (Fig. 9). Water availability is the most limiting factor for terrestrial carbon uptake.

At the same time, forests themselves strongly influence the availability of water. While there is little doubt that changes in tree cover will impact the water cycle, the wider consequences remain difficult to predict as the underlying relationships and processes remain poorly characterized.

In general, water transpiration of forests is much higher than of any other vegetation (and even than open water), typically exceeding transpiration of herbaceous vegetation by factor 10. The main reasons for this are that forest floors usually have a high water storage capacity; the deep root systems of trees that can mobilize water in lower soil layers and the water reservoirs in the stems allow transpiration to continue even under dry conditions.

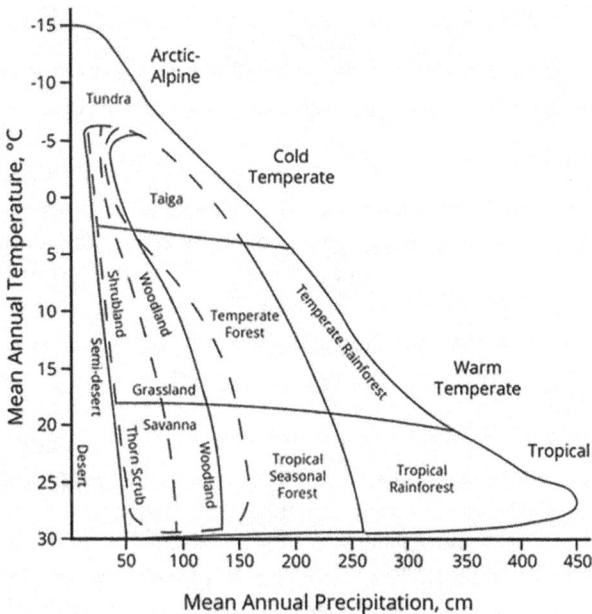

**Figure 9** Biome classification based on mean annual temperature and precipitation (modified from Whittaker, 1975).

Locally, increased transpiration into atmospheric moisture through forests can lead to reduced availability of water. Due to its increased evapotranspiration use, dense tree cover will reduce water availability compared with low tree cover. A meta-analysis found that dense plantations of *Eucalyptus* and *Pinus* reduced annual stream flow by over one-third, with larger losses at drier sites (Jackson et al., 2005).

Comparing only dense forests with bare land ignores the positive effects that an intermediately dense tree cover can have on soil water infiltration and soil desiccation. Studies in African drylands have shown that the positive effect of trees on water infiltration can greatly exceed evaporation losses and substantially increase, rather than deplete, ground water compared to treeless areas (Ilstedt et al., 2016). Furthermore, while local water yields often decline under young regrowth forest, water yields may recover as the vegetation matures (Filoso et al., 2017)

At a regional scale, however, the transpiration by forests and the atmospheric vapour they provide are of great importance. About 60% of the water in terrestrial precipitation is derived from the land itself (Schneider et al., 2017), and more than half of this land-derived atmospheric moisture comes from transpiration by plants (e.g. Jasechko et al., 2013). The typical distances that moisture evaporated from land travels in the atmosphere before it falls to earth once more are in the order of 500–5000 km. The Amazon Basin, East Africa, Western North America and Central Eurasia are major sources of water vapour for areas downwind (van der Ent and Savenije, 2011).

Reducing forest cover correlates with some decline in rainfall and cloud cover. It is estimated that land-cover change has caused about a 5% reduction in global atmospheric moisture (Sterling et al., 2013).

Atmospheric water transport is, to a large extent, understood as the result of atmospheric circulation induced by temperature gradients (Halley's temperature-gradient theory). This theory explains the sea-to-land winds, when the land is warmer than the ocean and the condensation of the transported moisture and its precipitation, when the air rises and cools down. The same theory also explains the monsoon cycles as a result of the seasonal switch as land-ocean temperature gradients switch direction from warmer ocean to warmer land and back. In this theory, the transpiration by forests is understood as mainly a passive recycling of terrestrial precipitation.

However, it is questioned whether the temperature-gradient theory is sufficient to explain by itself atmospheric motion and moisture transport. For example, it is difficult to explain with the temperature-gradient theory alone the inner-continental forests. Recent research suggests that condensation is not just the result of atmospheric moisture transport due to temperature gradient, but that the atmospheric moisture itself influences atmospheric dynamics (Makarieva et al., 2013).

Whenever water vapour condenses in the atmosphere as a result of the rising and cooling down of air, energy is released that heats up the air again, reduces its pressure and accelerates air flow. Within this condensation-driven theory, the role of forests in the atmospheric circulation is active rather than passive. Increased evapotranspiration by forests leads to increased condensation, which accelerates air flows and draws in moist air from the ocean that finally leads to rainfall that surpasses local evapotranspiration. This process is called the 'biotic pump' (Fig. 10).

Many observations support the existence of this biotic pump. In Southern African drylands, for example, the delayed increase in leaf area that follows rainfall reinforces subsequent winds and precipitation such the 'the vegetation draws airflow toward itself in a self-sustaining way' (Chikoore and Jury, 2010). The biotic pump also explains how high rainfall is maintained in the continental interiors of the Amazon, Congo and Siberia, while in areas without forest the

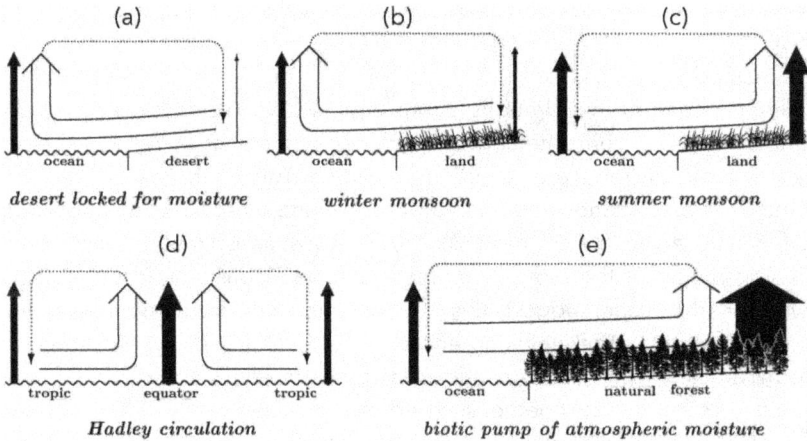

**Figure 10** Illustrations of the physical principle that the low-level air moves from areas with weak evaporation to areas with more intensive evaporation. Black arrows: evaporation flux, arrow width indicates the magnitude of this flux. Empty arrows: horizontal and ascending fluxes of moisture-laden air in the lower atmosphere. Dotted arrows: compensating horizontal and descending air-fluxes depleted of moisture in the upper atmosphere. (a) Deserts: evaporation on land is close to zero, so the low-level air moves from land to the ocean year-round, thus blocking the desert for moisture. (b) Winter monsoon: evaporation from the warmer oceanic surface is larger than evaporation from the colder land surface; the low-level air moves from land to the ocean. (c) Summer monsoon: evaporation from the warmer land surface is larger than evaporation from the colder oceanic surface; the low-level air moves from ocean to land. (d) Hadley circulation (trade winds): evaporation is more intensive on the equator, where the solar flux is stronger than in the higher latitudes. (e) Biotic pump of atmospheric moisture: transpiration fluxes regulated by natural forests exceed oceanic evaporation fluxes to the degree when the arising ocean-to-land fluxes of moist air become large enough to compensate losses of water to run off in the entire river basin year-round (Source: Makarieva and Gorshkov, 2006).

declines in rainfall with distance inland is steeper (typically halving each few hundred kilometres).

Temperature-driven atmospheric cycling and regional recycling not only explain these patterns, but also explain the 'Cold Amazon Paradox' where strongest winds from the Atlantic into the Amazon are observed when the Amazon is markedly colder than the ocean. The 'dry-season' transpiration typically permits some rainfall and this retained moisture facilitates the return of the wet season. Furthermore, temperature-based models are not able to explain abrupt shifts in monsoon climates found in paleo-data (Herzschuh et al., 2014).

If inland rainfall depends on large near-contiguous forest, then deforestation, especially near the coast, risks switching the continent from wet to dry. The precursor of such a switch would be declining rainfall and reduced reliability. This fits with patterns of declining rainfall in southwest Australia by 21% due to the reduction of native woody cover by 50% between 1950 and 1970 (Andrich and Imberger, 2013). It also fits with declining reliability in rainfall that correlates with forest loss on the Atlantic coast of Brazil (Webb et al., 2005).

Increasing desiccation and more frequent droughts can result in tree death, fires and further drought (Zemp et al., 2017). Furthermore, increased atmospheric $CO_2$ concentration that reduces stomatal opening and thereby transpiration will reduce atmospheric moisture. This might lead to reduced continental precipitation and related winds. On the other hand, the increasing leaf area observed in many regions may also boost transpiration.

The theory of the biotic pump, where forests constantly draw moisture away from the ocean, suggests this may also reduce the frequency and energy of tropical storms. The inhibiting effects of forests likely explain why cyclonic storms are rare in the South Atlantic. Loss of the Amazon and the Congo forests would increase the likelihood of such storms.

While a large area of planted forest will certainly influence the local water cycle, it is not yet clear whether it will be able to replace the climatic functions of natural forest. For example, compared with planted forests, many natural forests maintain lower understorey temperatures during the day and generate a local atmospheric inversion at night. Maintaining a moist understorey will influence how and when the water vapour is released. Tree diversity may also stabilize climate feedback by reducing vulnerability to changing conditions and disturbance.

## 6  Summary and future trends

Much progress has been made over the last few decades in understanding the role of forests in the global carbon cycle: the amount of $CO_2$ that is released to the atmosphere through deforestation and forest degradation (source) on one hand, and the amount of $CO_2$ sequestered by growing forests (sink) on the other. In parallel, political and financial incentives have been put in place in order to reduce emissions from deforestation and forest degradation as well as to promote sustainable management of forests and enhancement of forest carbon stocks through afforestation and forest restoration (REDD+).

Along with this, forest monitoring and carbon assessment capacities have been strengthened in many tropical countries over the last decade. At a scientific level, new remote-sensing instruments are being deployed that will permit monitoring of changes in global forest carbon stocks spatially explicitly, with ever higher precision as technology improves. NASA launched the GEDI satellite based on LiDAR technology in 2019 and ESA will launch the BIOMASS satellite based on P-Band radar in 2022.

With regard to the impact of forest management on climate change mitigation, it would be inadequate to look only at the $CO_2$ balance of the managed forest. Forest-based, long-term mitigation can only be achieved by a sustainable extraction of timber, its use for the replacement of materials and energy that depend on fossil fuels and finally the long-term storage of the $CO_2$ in the ground, either through biochar or BECCS technology.

The trade-off between maintenance or enhancement of forest carbon stocks and the increased use of wood needs to be analysed in more detail in order to provide clear guidance to decision-makers and forest managers. Technological development will be needed, particularly in tropical countries, for improving the efficiency of timber use (cascade use of wood), its end-use in energy generation and its long-term capture in the soil.

Compared with the $CO_2$ balance of forests, much less is known about the biophysical forest-climate interactions. There, the influence of forests on the water cycle and on atmospheric water transport is of particular importance. A very important question to address is the influence of forests on cloud formation and the feedback systems of increased water vapour and cloud coverage on the radiation budget. These interactions need to be understood in more detail to provide for a comprehensive assessment of climatic consequences of forest-cover change at the local, regional and global levels.

Our knowledge is very limited when it comes to the impact of climate change on the world's forests. Under which conditions can the negative climate – vegetation feedback (warmer climate -> more vegetation -> cooler climate) be maintained? At what point will the stress resulting from climate change turn forests into a net source of $CO_2$, turning the negative feedback into a positive feedback loop (warmer climate -> less vegetation -> warmer climate)? Again, these questions are strongly linked with the availability of water, as this is the most important limiting factor for forest growth, and whether $CO_2$ fertilization and increased water-use efficiency can compensate for the effect of drought.

Therefore, it is important to better understand how forests themselves attract water and are able to maintain the climatic conditions they need for their growth. The theory of the biotic pump provides an explanation for non-linear and rapid switches of vegetation observed in the past. The indication that a continent or region that passes some threshold of forest loss might tip from a wet to a dry climate, or, the other way round that rainfall can be stabilized and regained by maintaining and restoring tree cover.

# 7 Where to look for further information

In recent years, various synthesis publications have appeared on the current understanding of forest-climate interactions and the influence of forests on the global water cycle:

- Land-climate interactions: IPCC Special Report on climate change, desertification, land degradation, sustainable land management, food security and greenhouse gas fluxes in terrestrial ecosystems (Jia et al., 2019).
- Forests in the carbon cycle: The terrestrial carbon sink (Keenan and Williams, 2018).
- Risk of abrupt changes: Environmental Tipping Points (Lenton, 2013).
- Forests in the hydrological cycle: Forests, atmospheric water and an uncertain future: the new biology of the global water cycle (Sheil, 2018).

Valuable resources and guidelines have been developed by various organisations to assist policy makers and practitioners in integrating climate and hydrological services into forest management decisions:

- Global Forest Watch (WRI, 2019) providing maps with tree cover loss and associated $CO_2$ emissions:
  https://www.globalforestwatch.org/.
- Forestry for a low-carbon future. FAO (2016):
  http://www.fao.org/3/a-i5857e.pdf.
- Climate Change Guidelines for Forest Managers. FAO (2013):
  http://www.fao.org/3/i3383e/i3383e.pdf.
- Climate change for forest policymakers. FAO (2018):
  http://www.fao.org/3/CA2309EN/ca2309en.pdf.
- Forests and water. FAO (2005):
  http://www.fao.org/3/i0410e/i0410e.pdf.
- Forest Landscape Restoration as a Key Component of Climate Change Mitigation and Adaptation. IUFRO (2015):
  https://www.iufro.org/fileadmin/material/publications/iufro-series/ws34.pdf.

- Forest and Water on a Changing Planet: Vulnerability, Adaptation and Governance Opportunities. IUFRO (2018): https://www.iufro.org/fileadmin/material/publications/iufro-series/ws3 8/ws38.pdf.
- Carbon Accounting. Best practices in approaches at various scales for forest landscape restoration. IUCN (2015): https://portals.iucn.org/library/sites/library/files/documents/Rep-2015 -013.pdf.
- Estimating the mitigation potential of forest landscape restoration. Practical guidance to strengthen global climate commitments. IUCN (2019): https://portals.iucn.org/library/sites/library/files/documents/2019-029 -En.pdf.

# 8 References

Andrich, M. A. and Imberger, J. (2013). The effect of land clearing on rainfall and fresh water resources in Western Australia: a multi-functional sustainability analysis. *Int. J. Sust. Dev. World Ecol.* 20(6):549–563.

Baccini, A., Walker, W., Carvalho, L., Farina, M., Sulla-Menashe, D. and Houghton, R. A. (2017). Tropical forests are a net carbon source based on aboveground measurements of gain and loss. *Science* 358(6360):230–234. doi:10.1126/science.aam5962.

Bastin, J. F., Finegold, Y., Garcia, C., Mollicone, D., Rezende, M., Routh, D., Zohner, C. M. and Crowther, T. W. (2019). The global tree restoration potential. *Science* 365(6448):76–79. doi:10.1126/science.aax0848.

Brang, P., Spathelf, P., Larsen, J. B., Bauhus, J., Bonč̌ina, A., Chauvin, C., Drossler, L., Garcia-Guemes, C., Heiri, C., Kerr, G., Lexer, M. J., Mason, B., Mohren, F., Muhlethaler, U., Nocentini, S. and Svoboda, M. (2014). Suitability of close-to-nature silviculture for adapting temperate European forests to climate change. *Forestry* 87(4):492–503. doi:10.1093/forestry/cpu018.

Chikoore, H. and Jury, M. R. (2010). Intraseasonal variability of satellite-derived rainfall and vegetation over southern Africa. *Earth Interact.* 14(3):1–26.

Ciais, P., Sabine, C., Bala, G., Bopp, L., Brovkin, V., Canadell, J., Chhabra, A., DeFries, R., Galloway, J., Heimann, M., Jones, C., Le Quéré, C., Myneni, R. B., Piao, S. and Thornton, P. (2013) Carbon and Other Biogeochemical Cycles. In: Climate Change 2013: The Physical Science Basis. Contribution of Working Group I to the Fifth Assessment Report of the Intergovernmental Panel on Climate Change. Cambridge University Press, Cambridge, UK and New York, NY.

Claussen, M. Holocene rapid land-cover changes – evidence and theory. In: Battarbee, R. W. and Binney, H. A. (Eds). *Natural Climate Variability and Global Warming*. Wiley Blackwell, Oxford, UK (2008). 232–253. doi:10.1002/9781444300932.ch9.

Domke, G., Brandon, A., Diaz-Lasco, R., Federici, S., Garcia-Apaza, E., Grassi, G., Gschwantner, T., Herold, M., Hirata, Y., Kasimir, A., Kinyanjui, M. J., Krisnawati, H., Lehtonen, A., Malimbwi, R. E., Niinistö, S., Ogle, S. M., Paul, T., Ravindranath, N. H., Rock, J., Sanquetta, C. R., Sanz-Sanchez, M. J., Vitullo, M., Wakelin, S. J. and Zhu, J. (2019) 2019 Refinement to the 2006 IPCC Guidelines for National Greenhouse Gas Inventories.

FAO (2001). Global Forest Resources Assessment 2000 Main Report: FAO Forestry Paper 140. FAO, Rome (data from http://www.fao.org/3/Y0900E/y0900e05.htm#P4_44).

Filoso, S., Bezerra, M. O., Weiss, K. C. B. and Palmer, M. A. (2017). Impacts of forest restoration on water yield: a systematic review. *PLoS ONE* 12(8):e0183210.

Friedlingstein, P., Meinshausen, M., Arora, V. K., Jones, C. D., Anav, A., Liddicoat, S. K. and Knutti, R. (2014). Uncertainties in CMIP5 climate projections due to carbon cycle feedbacks. *J. Clim.* 27(2):511–526. doi:10.1175/JCLI-D-12-00579.1.

Harris, N. L., Brown, S., Hagen, S. C., Saatchi, S. S., Petrova, S., Salas, W., Hansen, M. C., Potapov, P. V. and Lotsch, A. (2012). Baseline map of carbon emissions from deforestation in tropical regions. *Science* 336(6088):1573–1576. doi:10.1126/science.1217962.

Herzschuh, U., Borkowski, J., Schewe, J., Mischke, S. and Tian, F. (2014). Moisture-advection feedback supports strong early-to-mid Holocene monsoon climate on the eastern Tibetan Plateau as inferred from a pollen-based reconstruction. *Palaeogeogr. Palaeoclimatol. Palaeoecol.* 402:44–54. doi:10.1016/j.palaeo.2014.02.022.

Houghton, R. A. and Nassikas, A. A. (2017). Global and regional fluxes of carbon from land use and land cover change 1850–2015. *Global Biogeochem. Cycles* 31(3):456–472. doi:10.1002/2016GB005546.

Ilstedt, U., Bargués Tobella, A., Bazié, H. R., Bayala, J., Verbeeten, E., Nyberg, G., Sanou, J., Benegas, L., Murdiyarso, D., Laudon, H., Sheil, D. and Malmer, A. (2016). Intermediate tree cover can maximize groundwater recharge in the seasonally dry tropics. *Sci. Rep.* 6:21930. doi:10.1038/srep21930.

Jackson, R. B., Jobbágy, E. G., Avissar, R., Roy, S. B., Barrett, D. J., Cook, C. W., Farley, K. A., Le Maitre, D. C., McCarl, B. A. and Murray, B. C. (2005). Trading water for carbon with biological carbon sequestration. *Science* 310(5756):1944–1947.

Jasechko, S., Sharp, Z. D., Gibson, J. J., Birks, S. J., Yi, Y. and Fawcett, P. J. (2013). Terrestrial water fluxes dominated by transpiration. *Nature* 496(7445):347–350.

Jia, G., Shevliakova, E., Artaxo, P., De Noblet-Ducoudré, N., Houghton, R., House, J., Kitajima, K., Lennard, C., Popp, A., Sirin, A., Sukumar, R. and Verchot, L. (2019). Land-climate interactions. In: IPCC Special Report on Climate Change, Desertification, Land Degradation, Sustainable Land Management, Food Security, and Greenhouse Gas Fluxes in Terrestrial Ecosystems.

Keenan, T. F. and Williams, C. A. (2018). The terrestrial carbon sink. *Annu. Rev. Environ. Resour.* 43(1):219–243. doi:10.1146/annurev-environ-102017-030204.

Keenan, T. F., Prentice, I. C., Canadell, J. G., Williams, C. A., Wang, H., Raupach, M. and Collatz, G. J. (2016). Recent pause in the growth rate of atmospheric CO2 due to enhanced terrestrial carbon uptake. *Nat. Commun.* 7:13428. doi:10.1038/ncomms13428.

Lenton, T. M. (2013). Environmental tipping points. *Annu. Rev. Environ. Resour.* 38(1):1–29. doi:10.1146/annurev-environ-102511-084654.

Leskinen, P., Cardellini, G., González-García, S., Hurmekoski, E., Sathre, R., Seppälä, J., Smyth, C., Stern, T. and Verkerk, P. J. (2018). *Substitution Effects of Wood-Based Products in Climate Change Mitigation. From Science to Policy 7*. European Forest Institute.

Lovett, G. M., Cole, J. J. and Pace, M. L. (2006). Is net ecosystem production equal to ecosystem carbon accumulation? *Ecosystems* 9(1):152–155. doi:10.1007/s10021-005-0036-3.

Makarieva, A. and Gorshkov, V. (2006). Biotic pump of atmospheric moisture as driver of the hydrologicalcycle on land. *Hydrol. Earth Syst. Sci. Discuss.* 3(4):2621–2673. European Geosciences Union.

Makarieva, A. M., Gorshkov, V. G., Sheil, D., Nobre, A. D. and Li, B.-L. (2013). Where do winds come from? A new theory on how water vapor condensation influences atmospheric pressure and dynamics. *Atmos. Chem. Phys.* 13(2):1039–1056.

Nabuurs, G. J., Masera, O., Andrasko, K., Benitez-Ponce, P., Boer, R., Dutschke, M., Elsiddig, E., Ford-Robertson, J., Frumhoff, P., Karjalainen, T., Krankina, O., Kurz, W. A., Matsumoto, M., Oyhantcabal, W., Ravindranath, N. H., Sanz Sanchez, M. J. and Zhang, X. (2007) Forestry. In *Climate Change 2007: Mitigation. Contribution of Working Group III to the Fourth Assessment Report of the Intergovernmental Panel on Climate Change.* Cambridge University Press, Cambridge, UK and New York, NY.

Oyama, M. D. and Nobre, C. A. (2003). A new climate-vegetation equilibrium state for Tropical South America. *Geophys. Res. Lett.* 30. doi:10.1029/2003GL018600.

Pan, Y., Birdsey, R. A., Fang, J., Houghton, R., Kauppi, P. E., Kurz, W. A., Phillips, O. L., Shvidenko, A., Lewis, S. L., Canadell, J. G., Ciais, P., Jackson, R. B., Pacala, S. W., McGuire, A. D., Piao, S., Rautiainen, A., Sitch, S. and Hayes, D. (2011). A large and persistent carbon sink in the world's forests. *Science* 333(6045):988–993. doi:10.1126/science.1201609.

Potapov, P., Laestadius, L. and Minnemeyer, S. (2011). *Global Map of Forest Landscape Restoration Opportunities.* World Resources Institute, Washington, DC. Available at: www.wri.org/forest-restoration-atlas.

Scheffer, M., Carpenter, S. R., Foley, J. A., Folke, C. and Walker, B. H. (2001). Catastrophic shifts in ecosystems. *Nature* 431:591–596.

Schneider, U., Finger, P., Meyer-Christoffer, A., Rustemeier, E., Ziese, M. and Becker, A. (2017). Evaluating the hydrological cycle over land using the newly-corrected precipitation climatology from the global precipitation climatology Centre (GPCC). *Atmosphere* 8(12):1–17.

Sheil, D. (2018). Forests, atmospheric water and an uncertain future: the new biology of the global water cycle. *For. Ecosyst.* 5(1):19. doi:10.1186/s40663-018-0138-y.

Song, X. P., Hansen, M. C., Stehman, S. V., Potapov, P. V., Tyukavina, A., Vermote, E. F. and Townshend, J. R. (2018). Global land change 1982-2016. *Nature* 560(7720):639–643. doi:10.1038/s41586-018-0411-9.

Sterling, S. M., Ducharne, A. and Polcher, J. (2013). The impact of global land-cover change on the terrestrial water cycle. *Nat. Clim. Chang.* 3(4):385–390.

Taverna, R., Hofer, P., Werner, F., Kaufmann, E. and Thürig, E. (2007). *CO2-Effekte der Schweizer Wald- und Holzwirtschaft. Szenarien zukünftiger Beiträge zum Klimaschutz. Umwelt-Wissen Nr. 0739.* Bundesamt für Umwelt, Bern, 102S.

van der Ent, R. J. and Savenije, H. H. G. (2011). Length and time scales of atmospheric moisture recycling. *Atmos. Chem. Phys.* 11(5):1853–1863.

Webb, T. J., Woodward, F. I., Hannah, L. and Gaston, K. J. (2005). Forest cover-rainfall relationships in a biodiversity hotspot: the Atlantic Forest of Brazil. *Ecol. Appl.* 15(6):1968–1983. doi:10.1890/04-1675.

Whittaker, R. H. (1975). *Communities and Ecosystems.* Macmillan Publishing Company, New York, NY.

Zemp, D. C., Schleussner, C. F., Barbosa, H. M. J., Hirota, M., Montade, V., Sampaio, G., Staal, A., Wang-Erlandsson, L. and Rammig, A. (2017). Self-amplified Amazon forest loss due to vegetation-atmosphere feedbacks. *Nat. Commun.* 8:14681. doi:10.1038/ncomms14681.

Zhu, Z., Piao, S., Myneni, R. B., Huang, M., Zeng, Z., Canadell, J. G., Ciais, P., Sitch, S., Friedlingstein, P., Arneth, A., Cao, C., Cheng, L., Kato, E., Koven, C., Li, Y., Lian, X., Liu, Y., Liu, R., Mao, J., Pan, Y., Peng, S., Peñuelas, J., Poulter, B., Pugh, T. A. M., Stocker, B. D., Viovy, N., Wang, X., Wang, Y., Xiao, Z., Yang, H., Zaehle, S. and Zeng, N. (2016). Greening of the earth and its drivers. *Nat. Clim. Change* 6(8):791–795. doi:10.1038/nclimate3004.

# Chapter 4

## Forest landscape restoration (FLR) of tropical forests

*Stephanie Mansourian, Mansourian.org/University of Geneva, Switzerland/IUFRO, Austria*

## 1  Introduction

### 1.1 Forest loss and degradation

While 5000 years ago forests covered around 5.8 billion ha of the planet, today they cover just below 4 billion ha (FAO, 2016). According to the United Nations Food and Agriculture Organization we are losing 7.6 million ha of forest annually (FAO, 2015). An even larger area of forests is being degraded each year although degradation is harder to quantify as what counts as degradation for one person (e.g. a eucalyptus plantation) may be seen as restored land by another person (Sasaki and Putz, 2009; Thompson et al., 2013). Nevertheless, in 2002, the International Tropical Timber Organization (ITTO) estimated forest degradation to have affected up to 850 million ha in tropical regions alone (ITTO, 2002). The loss and degradation of forests have an impact on the delivery of a wide range of goods and services provided by forests, such as pollination, biodiversity, medicines, carbon sequestration, water regulation, soil improvement, to cite but a few (IPBES, 2018).

Faced with this decline of forest quality and quantity, there has been a growing recognition of the urgency to restore forests (e.g. Aronson and Alexander, 2013). Governments have, for many decades, pledged to plant large numbers of trees, with, for example, Algeria having an 80 000 ha reforestation target after independence from France in the 1960s (Bensaid, 1995). More recently, in 1998, Vietnam launched its '5 million ha reforestation programme',

http://dx.doi.org/10.19103/AS.2020.0074.38

and in 2011, several governments rallied around the Bonn Challenge on forest landscape restoration (FLR) and committed to restoring 150 million ha by 2020 and 350 million ha by 2030. Many private sector initiatives also exist, including the 2020 World Economic Forum (WEF) '1 trillion trees' initiative (1t.org).

## 1.2 Restoring forests

Reversing forest loss and degradation may be possible through many different means. Trees can be planted, but forest areas may also be protected to enable them to regenerate naturally. The species of trees selected and the ways in which they are planted may also vary. Soil and land preparation techniques differ. Policies and incentives can encourage or enforce restoration. Many different terms, such as afforestation, reforestation, rehabilitation and ecological restoration, are associated with this reversal of forest cover. Mansourian (2018) documented 24 different terms to denote the different ways in which forests can be re-established. Forest landscape restoration (FLR) is one such term. Its roots can be traced back to forestry and to the conservation community, particularly in restoration ecology and conservation biology.

## 1.3 At the root of FLR

Traditionally, forestry practices have focussed on managing forests for the delivery of timber. In many countries to this day, the forestry departments work with a limited number of species and singular objectives. Nevertheless, amid growing criticism, forestry departments in many countries - such as Switzerland and the United States - have adopted more complex, multi-purpose approaches to their forestry practices (Wiersum, 1995). The selection of species to be planted in these more comprehensive approaches has been closer to the native structure, more diverse and with various objectives, such as soil stabilisation, avalanche and water protection, in addition to the traditional provision of timber (Lamb and Gilmour, 2003).

The discipline of restoration ecology is relatively recent with the scientific body, the Society for Ecological Restoration (SER) established in 1998 in the United States. Ecological restoration is defined as 'an intentional activity that initiates or accelerates the recovery of an ecosystem[1] with respect to its health, integrity and sustainability' (Clewell et al., 2004). The science of restoration ecology has generally focussed on small sites, and has based its approach on reference ecosystems - sites that contain the assumed 'historical or original ecosystem' - as

---

1 Ecosystem is defined by the CBD as 'a dynamic complex of plant, animal and micro-organism communities and their non-living environment interacting as a functional unit' (CBD, 1992).

a model for determining actions. The emphasis is on the use of native vegetation, natural regeneration and facilitating natural processes, among others (e.g. Clewell et al., 2005; Hallett et al., 2013; Allison and Murphy, 2017). More recently, restoration ecology has taken a more pragmatic approach, acknowledging that reference ecosystems merely provide guidance and direction, that larger scales are important, and that the human dimension also needs to be taken into account (Holl et al., 2003; Perring et al., 2018; Gann et al., 2019).

For conservation biologists, traditionally, their main implementation tool has been protected areas (e.g. Margules and Pressey, 2000; Brooks et al., 2004). A gradual shift to larger areas – ecoregions and hotspots – has broadened the scope of interventions (Olson and Dinerstein, 1998; Myers et al., 2000). This coincided with improved technological tools to support the mapping of these larger areas, in particular, geographic information systems (GIS) (Redford et al., 2003). The landscape was perceived as a more practical scale, somewhere between an ecoregion and a site. In these larger contexts, new techniques were required, notably restoration to improve connectivity and habitat functionality (e.g. Bennett, 1999). Upscaling also brought about the need for conservationists to better integrate humans into their strategies. Efforts were made to promote the co-management of natural resources, including forests, whereby local communities were able to share the benefits of improved conservation. Such initiatives began in the 1990s in sub-Saharan Africa (Roe et al., 2009) and even earlier in South Asia (Agrawal and Chhatre, 2006). Integrated conservation and development projects (ICDPs) were developed in the late 1990s as an attempt to better align social and ecological objectives (McShane and Wells, 2004).

FLR emerged as a complex melting pot of the integration of forestry practices, restoration ecology, conservation biology and ICDPs. It was defined in 2000 by a group of 30 experts convened by WWF and IUCN as 'a planned process that aims to regain ecological integrity and enhance human wellbeing in deforested or degraded landscapes' (WWF and IUCN, 2000). In practice, FLR does not necessarily require the return of trees across a whole landscape, but rather the strategic location of trees and forests in the landscape to ensure that they can maximise the benefits that forests provide to people and biodiversity. The original intention through this approach was three-pronged: (1) to raise the spatial scale at which restoration was being carried out, (2) to marry both ecological and human dimensions and, therefore, necessarily address multiple objectives and (3) to acknowledge that restoration is a (long-term) process. The term 'planned' was included in the definition in recognition of the fact that, although some adaptation may be required, the desire to reach a given objective necessarily requires a plan for implementation. Since then, many adaptations of this definition have been in use; notably some have removed the word 'planned' implying a more dynamic evolution of the process over time. The 'Global Partnership on FLR' (GPFLR) formally established in 2003 by

IUCN, WWF and the UK Forestry Commission, and which now re-groups 30 governments and international organisations, uses the definition 'a process that aims to regain ecological functionality and enhance human well-being in deforested or degraded landscapes'. The shift to the term 'ecological functionality' rather than 'ecological integrity' also reflects, on the one hand, the complexity of the term 'ecological integrity' and, on the other hand, the growing emphasis on forest functions, notably carbon sequestration (Pistorious and Kiff, 2017).

In part to address the diverging definitions and approaches to FLR, the GPFLR defined six principles for FLR in 2018: (1) focus on landscapes; (2) engage stakeholders and support participatory governance; (3) restore multiple functions for multiple benefits; (4) maintain and enhance natural ecosystems within landscapes; (5) tailor to the local context using a variety of approaches; and (6) manage adaptively for long-term resilience (Besseau et al., 2018).

### 1.4 From practice to politics

Although originating in the environmental conservation community, FLR became a political process when in 2011 the government of Germany, together with IUCN, launched the Bonn Challenge. Intended to raise awareness about the urgent need for restoration, and about the value of FLR to address multiple objectives, the Bonn Challenge rapidly became a banner for the promotion and widespread adoption of FLR and tree planting more generally.

With the ongoing loss and degradation of forests, it is evident that the restoration of forests is required. However, how this happens, who takes decisions about where, what and how to restore, who is involved, who benefits and who loses, are all questions that remain often unanswered or inadequately addressed (Boedhihartono and Sayer, 2012; Mansourian and Sgard, 2019). Trade-offs are generated by the desire to address both human and ecological objectives, which may not always be compatible (Boedhihartono and Sayer, 2012). For example, fencing-off an area to allow natural regeneration may bar access to key products such as fuelwood or food products to local communities. Unless the underlying causes of forest loss and degradation are addressed, FLR will also fail to be implemented effectively (Stanturf et al., 2019).

## 2 Implementing forest landscape restoration (FLR)

Technical interventions for FLR include understanding the soil properties, genetics, biodiversity, forestry dynamics, and so on. Implementing FLR is not just a technical matter however; additional factors to consider include tenure, policies, funding, and cultural factors (see Table 1). Furthermore, in practice,

**Table 1** Key areas of intervention (reproduced from Sabogal et al., 2015)

| Key areas of intervention | Factors to take into account |
|---|---|
| Assessment of landscape degradation and restoration opportunities | • Decide on the most appropriate assessment methodologies to use (e.g. ROAM).<br>• Identify degraded lands and the best opportunities for successful restoration efforts.<br>• Identify main agents and drivers of degradation within landscapes.<br>• Assess the ecological conditions, social-cultural dynamics and other enabling factors.<br>• Carry out stocktaking of successful interventions.<br>• Analyse and evaluate costs and benefits of selected restoration options and carry out a risk assessment of those options for investors. |
| Enabling environment | • Analyse policies, laws and regulations across different sectors. Are they adequate? Are they complementary/conflicting?<br>• Support drafting, revision and/or harmonization of laws/policies/sectoral programmes and identify specific support, activities and projects to create a more enabling environment. |
| Institutional setting | • Identify relevant land-use sectors and stakeholders for FLR (forestry, agriculture, livestock/rangeland, energy, mining, etc.).<br>• Support planning processes that are underway (e.g. climate-change national strategy, biodiversity national strategy, national strategy for rural development, etc.).<br>• Consider all relevant entry points as FLR can be an effective package to generate and share a range of benefits (e.g. biodiversity, food security, climate mitigation and adaptation, livelihoods, poverty alleviation, etc.).<br>• Identify/support existing mechanisms/platforms that allow different sectors/stakeholders to engage in dialogue.<br>• Identify and leverage existing partnerships. |
| Governance issues | • Assess land-tenure issues and try to secure tenure, especially for local stakeholders, to allow investments in FLR.<br>• Identify barriers to people's participation.<br>• Analyse decision-making processes.<br>• Facilitate engagement of all relevant stakeholder groups. |
| Technologies and approaches | • Carry out stocktaking of existing technologies and approaches for sustainable land use (reforestation, assisted natural regeneration, agroforestry, climate-smart agriculture, agroecology, etc.).<br>• Build on successful experiences and approaches already carried out.<br>• Set up a portfolio of cost-effective and ecologically robust restoration techniques.<br>• Identify knowledge gaps. |

*(Continued)*

**Table 1** (*Continued*)

| Key areas of intervention | Factors to take into account |
|---|---|
| Capacity development and extension | • Identify capacity-development needs at the individual and organizational level and propose relevant strategies to meet these needs.<br>• Build capacity-development programmes for relevant stakeholders to undertake planning, implementation and evaluation of FLR efforts.<br>• Develop networks/knowledge platforms (national/regional) between practitioners and extension services in order to disseminate good practices.<br>• Support the establishment and continued capacity strengthening of networks of practitioners and extension services. |
| Resource mobilization | • Estimate the resources already available through existing national/subnational programmes/projects.<br>• Elaborate national action plans or national strategies as the basis for building trust with donors in terms of national commitment to FLR.<br>• Integrate FLR into state budgets and public investment funds.<br>• Develop monitoring systems for FLR expenditures and mechanisms for collecting data on the costs and benefits of FLR.<br>• Devise a coordinated approach to informing/sensitizing potential donors (multilateral, bilateral, foundations, etc.) and support the development of new project proposals.<br>• Mobilize innovative sources of funding through mechanisms such as climate finance instruments and/or payment for environmental services and develop incentive packages that include economic and non-economic benefits.<br>• Design, adapt and implement national and local financing mechanisms for FLR, in particular by promoting the development of financial instruments at the local level (e.g. local development funds, microfinance instruments, credit lines in local private banks), with positive incentives for local stakeholders to promote sustainable FLR investments.<br>• Use these financing instruments to implement public incentive schemes (e.g. payments for ecosystem services) and couple these schemes with investments in sustainable value chains to ensure a long-term, self-sustaining financing strategy. |
| Private sector investment | • Increase engagement with the private sector, especially with pioneer private-impact funds and other innovative initiative funds as key partners in the FLR investment continuum.<br>• Understand the scope of private-sector involvement in FLR already underway in the country and build a legal and regulatory framework that promotes landscape 'readiness for investments' and attracts investors to FLR.<br>• Facilitate the dialogue between the private sector and other stakeholders in order to decrease transaction costs for private-sector investments.<br>• Develop a pipeline of bankable restoration projects and raise the awareness of the private sector about FLR opportunities in key value chains (marketplace).<br>• Foster favourable conditions for public–private partnerships and promote risk-mitigation mechanisms to engage FLR investors at scale. |

| Key areas of intervention | Factors to take into account |
|---|---|
| Information dissemination and research needs | • Facilitate regular access to relevant information with practical knowledge and experiences targeting varied audiences.<br><br>• Identify (biophysical, socioeconomic, etc.) gaps in knowledge that research institutions could address more effectively. Emphasize research geared to innovative solutions for local stakeholders.<br><br>• Develop robust indicators adapted to the local/national context and develop consistent monitoring systems in order to improve theeffectiveness of FLR efforts. |

implementation may have different temporal and spatial scales, as well as multiple objectives (Sabogal et al., 2015). Challenges with implementing FLR emerge because of its long-term and iterative nature, therefore requiring adaptive management as circumstances change. Furthermore, FLR is context-specific, so there is a limit to how much standardised guidance can be provided. It requires a balance between planning at a larger (landscape) scale and implementation at a local (site) scale.

## 2.1 Implementation guidance

Different implementation guides have been designed for FLR. Chazdon and Guariguata (2018) identify a number of decision-support tools for FLR, some of which are highlighted here. For Vallauri et al. (2005) the main steps in FLR are: (1) initiating a restoration programme and partnerships (identifying problems and agreeing on solutions); (2) defining restoration needs and linking restoration to a large-scale conservation vision; (3) defining restoration strategy and tactics, including land-use scenarios; (4) implementing restoration; and (5) piloting systems towards fully restored ecosystems (including evaluation and corrective actions).

The World Resources Institute (WRI) identifies six steps to upscaling regreening in the framework of FLR (Reij and Winterbottom, 2015): (1) identify and analyse existing regreening successes; (2) build a grassroots movement for regreening; (3) address policy and legal issues and improve enabling conditions for regreening; (4) develop and implement a communication strategy; (5) develop or strengthen agroforestry value chains and capitalise on the role of the market in upscaling regreening; and (6) expand research activities to fill gaps in knowledge about regreening.

The AFR100 initiative identifies three broad stages for FLR implementation: (1) start – by sharing information, identifying key actors, finding successful cases, diagnosing barriers to upscaling and mapping restoration activities; (2) implement – by developing a strategy for implementation, improving enabling conditions, committing to targets, integrating FLR into policies and developing monitoring methods and (3) scale – establishing baselines, tracking policy reforms, monitoring adherence to standards, sharing lessons learnt and instituting periodic reviews (NEPAD, 2017).

More recently, applying the basic steps in project cycle management Stanturf et al. (2019) highlighted four stages for FLR implementation: (1) visioning – where goals for FLR are identified; (2) conceptualising – where goals are broken down into specific and measurable objectives; (3) acting – where specific activities are defined and an action plan with clear responsibilities established, and (4) sustaining – where the need for long-term monitoring and adaptive management is acknowledged.

All of these guides have in common four key phases in FLR implementation. First, they identify a design phase which relates to engaging stakeholders and partners and agreeing on a vision for the future landscape. Second, they highlight a planning phase during which goals and objectives are defined, relevant policies are targeted, funds and tools identified and so on. The next phase is the implementation one, where actions are taken at different scales, and tools are applied. Finally, the monitoring and adapting phase uses feedback to inform future phases of the process.

Concrete interventions are manifold (see Table 1). They may occur in the forestry sector (e.g. identifying quality seeds and preparing soils), but they may also occur at the political level (e.g. defining policies that encourage restoration). Indeed, policies, incentives, institutions and stakeholder platforms are all critical to the ultimate realisation of FLR. Seven types of 'FLR interventions' are highlighted in the IUCN-WRI Restoration Opportunities Assessment Methodology (ROAM): (1) planted forests and woodlots; (2) natural regeneration; (3) silviculture; (4) agroforestry; (5) improved fallow; (6) mangrove restoration and (7) watershed protection and erosion control. The definition of specific, locally appropriate activities will depend on many factors, notably the climatic zones, the socio-political conditions, the time since deforestation/degradation occurred, the level of investment, etc. For example, in a densely populated area where deforestation and forest degradation have been rampant because of illegal logging and small-scale farming, it may be more appropriate to consider reinforcing policies to limit further degradation and forest loss, while also working with local farmers to include trees on farms. In another case where the removal of riparian forests has affected watercourses, incentives may be put in place by the government to support farmers and landowners to restore these valuable forests, while distributing free seedlings of native species to support the restoration effort.

In all cases, FLR is a long-term and iterative process, with feedback loops at all stages. Much can change over the time frames (measured in decades) of FLR interventions, and adaptive management is essential (Larson et al., 2013; Dudley et al., 2018).

## 2.2 Tools

Specific tools for FLR have been developed by different organisations. Tools have generally focussed on forestry interventions, with only a few to date attempting to be more comprehensive. For example, the WRI and IUCN in 2014 developed the ROAM, which is a way of analysing and defining the package of opportunities for FLR in a given country or landscape (IUCN and WRI, 2014). The World Resources Institute (WRI) has developed a 'restoration diagnostic' to define strategies for FLR (Hanson et al., 2015). The International Union of Forest Research Organizations (IUFRO) developed an implementation guide that provides some guidance on the different aspects to consider when implementing FLR (Stanturf et al., 2017). It takes the reader through the main stages of FLR, using examples and a checklist under each module to support the user. For example, Module II highlights some of the governance aspects to consider in FLR such as identifying stakeholders, their roles and interactions, as well as tenure, conflict resolution and negotiations. Module IV on technical interventions provides a decision tree to help the user determine which specific forestry intervention makes sense under given conditions. The WWF has developed a series of reports on the lessons learnt from its various FLR initiatives (published to date on Madagascar, New Caledonia, Tanzania, the Lower Danube, Mexico and Borneo, with the Atlantic Forest in the pipeline) which provide a comprehensive retrospective of lessons learnt from some of the first FLR projects designed (see website: http://forestsolutions.panda.org/approach/flr). At Yale University, the Environmental Leadership and Training Initiative (ELTI) provides a capacity-building programme on FLR with both field and online activities (see: https://elti.yale.edu/).

## 3 Case studies

### 3.1 Case study 1: WWF in Madagascar's Fandriana-Marolambo landscape

In 2004, the international non-governmental organization WWF raised funds from the French government to carry out a 3-year project on FLR in Madagascar. The Fandriana-Marolambo (FM) landscape located in east-central Madagascar was identified as a priority landscape for this project. It spans an area of about 200 000 ha and is primarily composed of a mosaic of primary forest, degraded primary forest, secondary forest, savannah and agricultural areas. An estimated 150 000 people live in the landscape representing three different ethnic groups: Betsileo, Vakinankaratra and Betsimisaraka (Mansourian et al., 2016).

The project objective was that 'the goods, services and authenticity of the moist forests of the landscape of Fandriana-Marolambo are restored so as to support the development of the populations and to secure the objectives of biodiversity conservation' (Mansourian et al., 2018). This ambitious objective could not be realistically achieved in the project time frame, and a further three projects were designed as follow-up phases (Ibid). Overall a total of approximately €1.6 million was spent in the landscape for FLR over a 13-year period (2005–2018). Interventions included: agreeing on a joint vision for the landscape; clarifying land status and tenure; introducing the FLR concept and ensuring it was adopted within local development plans; building on local associations; supporting and initiating long-term sustainability; adapting agricultural practices; nursery development; transferring management rights to local communities; identifying and carrying out alternative income-generating activities with farmers such as improved rice production, the production of essential oils, honey, and small animal and fish farming.

A total of 50 locally run nurseries, growing 100 native species were established, and 6 786 ha placed under active or passive restoration. One challenge in the landscape lay in obtaining acceptance of external intervention by local communities without whom nothing could be achieved. Also, because of poverty levels and reliance by local communities on fuelwood, trade-offs in the landscape had to be considered, including mixing fast-growing eucalyptus trees (for fuelwood needs of the local communities) with native trees. Highlighted lessons include the importance of monitoring, capacity building and multilevel partnerships.

In practice, the project has raised awareness and mobilised many actors. However, after 13 years of implementation, the impact on the ground is limited, demonstrating the complexity of such multipronged approaches, in large areas, and with a diversity of stakeholders, facing many basic livelihood challenges. The government of Madagascar has committed to restoring 4 million ha by 2030 under the Bonn Challenge, and lessons from this project may provide useful insights into how to achieve such an ambitious target.

### 3.2 Case study 2: Bonn Challenge on FLR

At an international level, the political process termed as the 'Bonn Challenge on FLR' represents a different level of intervention from the previous case. It is a demonstration of political mobilisation around a particular land-use intervention such as FLR. Launched in 2011, the Bonn Challenge set out to restore 150 million ha by 2020 and 350 million ha by 2030. As of March 2020, the Bonn Challenge had 172.35 million ha pledged from 62 national and regional governments, as well as one company. Although voluntary, the Bonn Challenge is seen as a means of supporting governments' actions under the three main environmental conventions (the UNFCCC, the UNCCD and the CBD) tackling at the same time climate change (through carbon sequestration), degradation (through soil retention, improved land cover) and biodiversity (through increased and improved habitat).

The Bonn Challenge also spawned a series of regional initiatives: the AFR 100 that aims to bring 100 million ha into restoration in Africa by 2030, the Initiative 20x20 in Latin America that aims to bring 20 million ha into restoration in this region by 2020, and the ECCA30 that seeks to bring 30 million ha of degraded and deforested land in Europe, the Caucasus and Central Asia into restoration by 2030. In recognition of the long-term process required for restoration to take place, the wording has evolved under these political platforms to reflect that the aim is to bring areas 'under restoration' rather than explicitly refer to 'areas restored'.

While it has provided a launch pad for FLR, the Bonn Challenge has also highlighted the frequent disconnect between policy and practice (and science). In fact, the strong political mobilisation around FLR has been both a challenge and an opportunity for wider FLR interventions. On the one hand, it has provided an entry point for FLR interventions and helped to raise funding for these. On the other hand, it has put pressure to achieve quick results, with a very limited understanding of the complexity of the issues, the time frames involved and the need for multiple level interventions (Mansourian et al., 2017b; Perring et al., 2018).

## 4  Challenges and opportunities in taking FLR forward

In 2017, reflecting on 12 years of implementation of FLR, Mansourian et al. (2017a) noted that the main challenges still affecting FLR include: (1) implementing FLR at scale and in an interdisciplinary fashion, bringing together multidisciplinary teams; (2) improving governance, particularly considering important issues of direct relevance to FLR such as tenure; (3) ensuring an increase in both forest cover and forest quality, rather than just focussing on quantity; (4) promoting the role of restoration in climate change responses, acknowledging that carbon sequestration is just one of the multiple objectives one can set for FLR in any given area; and (5) improving methodologies for measuring long-term impacts and their application.

Although FLR has evolved into a major land-use option, evidence of its effective and widespread implementation remains scant (Mansourian et al., 2017a). Of the many challenges confronting FLR, turning political will into practical action is probably the first and foremost. The areas committed under the Bonn Challenge are ambitious. While the global challenges of biodiversity loss and climate change require significant action (including large-scale restoration) in practice, finding areas suitable for FLR and engaging for such long-term action is difficult. Increasingly, the land is at a premium and under pressure for urban expansion, infrastructure and food production (IPBES, 2018). The risk of green grabbing whereby land that is of importance for local communities is appropriated for FLR is a concern (Lund et al., 2017).

Highly visible initiatives, such as the Bonn Challenge, risk highlighting quick results, at the expense of the complex mix of interventions required over the long term to deliver sustained and sustainable benefits. Simple solutions that do not consider local contexts and are not tailor-made and adapted to the socioecological conditions (Pandit et al., 2020) may be favoured for expediency. In this way, the political hype around FLR may well backfire. Seeking to maintain quality interventions is a real concern.

Climate change represents both an opportunity and a challenge for FLR. On the one hand, forests contribute to climate change when they burn; on the other, they can help to reduce it through their absorption of carbon as they grow (Mansourian et al., 2017a). Planting trees is, therefore, a recognised and widespread action to reduce and offset greenhouse gas emissions. For example, many companies include these actions in their carbon-offsetting strategies. In practice, FLR may also contribute to reducing carbon emissions as well as providing many additional long-term benefits such as the provision of non-timber forest products to local communities, and services in the landscape such as pollination, soil retention, wind protection, improving the microclimate, and so on. The risk is that under the guise of carbon offsetting, inappropriate strategies are used (notably large-scale plantations of single species, on land that may be contested by different groups). Much uncertainty exists as well concerning the vulnerability of forests to climate change and, therefore, their ability to effectively contribute to its mitigation (Anderson-Teixeira et al., 2013).

Annual financing needs for FLR globally have been estimated from $36 billion to $49 billion (FAO and Global Mechanism of the UNCCD, 2015). This represents about twice the total global funding for the environment under the Global Environment Facility since its creation in 1991 (GEF website). This staggering amount can only be met through multiple funding sources, including in particular the private sector (Löfqvist and Ghazoul, 2019). Recognising (and monetising) the benefits that FLR can provide, as well as the savings through improved resilience, adaptation and marketable ecosystem services (such as freshwater provision) can also contribute to more realistic calculations of funding needs (TEEB, 2009).

Discussing and negotiating necessary trade-offs in the landscape (both social and ecological), which is often necessary for FLR implementation, is tricky. Inequalities among different stakeholders, imbalanced power relations and the lack of tools to effectively engage and negotiate with stakeholders hamper this essential element of FLR (Mansourian et al., 2017a). More broadly, understanding the governance of FLR – that is, who takes decisions related to FLR – remains woefully inadequate, and much research is required in this area (Mansourian, 2016).

A lack of integration at many levels has been identified as a challenge for FLR (Mansourian and Parrotta, 2019). This includes lack of integration across spatial scales (from local to international), across sectors (e.g. agriculture and forestry), across knowledge systems (e.g. indigenous knowledge and western knowledge), between social and ecological systems, across timescales (from the past to the present and the future) and between formal and informal governance systems.

Monitoring restoration of forested landscapes requires both social and ecological dimensions. In practice, monitoring is often neglected in many conservation projects, and FLR is no different. In particular, given the complexity of issues to be addressed in FLR, it remains challenging to define, design and apply simple measures to monitor FLR progress (Dudley et al., 2018; Pandit et al., 2020).

Having said that, FLR presents major opportunities. It represents a pragmatic approach to meet multiple challenges related to land use. The multiple and fundamental roles that forests play in landscapes are well recognised, spanning all four categories of ecosystem services identified by the Millennium Ecosystem Assessment (supporting, provisioning, regulating and cultural) (MEA, 2005). FLR can contribute to all of these services. Because FLR does not necessarily imply large scale tree planting, but rather acknowledges the need for trade-offs in the landscape, it is more likely to be accepted by multiple stakeholders present in the landscape. By considering both ecological and human dimensions, it acknowledges that a forest landscape is a social-ecological system (Ostrom, 2009; Yang et al., 2018). The widespread global mobilisation on FLR, led by the Bonn Challenge and the political impetus it has generated, provides an effective launch pad for resource mobilisation, research and implementation of FLR initiatives.

## 5 Conclusion

Defined 20 years ago, FLR has evolved from a conservation-led approach to a global political movement. It has the potential to offer a solution to many global environmental issues, but its widespread implementation remains a challenge. While there are many efforts and initiatives towards FLR, many would not strictly comply with the principles of FLR, and others struggle to reach the scale intended. Nevertheless, restoration is a long-term process, and seeds planted in this decade will take at least one generation to turn into forests. Continued efforts are necessary to improve the science and practice of FLR and create the right political, social and economic conditions for it to be a viable long-term solution to deforestation, land and forest degradation, and related loss of ecosystem goods and services.

## 6   Where to look for further information

Major organisations involved in this topic are IUCN (iucn.org and infoflr.org); WRI (wri.org); IUFRO (iufro.org); FAO (fao.org) and WWF (panda.org).

The Global Partnership on FLR which re-groups over 30 actors has its own website at: http://www.forestlandscaperestoration.org/.

CIFOR (cifor.org) has also published some specific research on the topic, for example, on gender and FLR.

## 7   References

Agrawal, A. and Chhatre, A. 2006. Explaining success on the commons: community forest governance in the Indian Himalaya. *World Development* 34(1), 149-66. doi:10.1016/j.worlddev.2005.07.013.

Allison, S. K. and Murphy, S. D. (Eds) 2017. *Routledge Handbook of Ecological and Environmental Restoration*. Routledge, Oxon.

Anderson-Teixeira, K. J., Miller, A. D., Mohan, J. E., Hudiburg, T. W., Duval, B. D. and DeLucia, E. H. 2013. Altered dynamics of forest recovery under a changing climate. *Global Change Biology* 19(7), 2001-21. doi:10.1111/gcb.12194.

Aronson, J. and Alexander, S. 2013. Ecosystem restoration is now a global priority: time to roll up our sleeves. *Restoration Ecology* 21(3), 293-6. doi:10.1111/rec.12011.

Bennett, A. F. 1999. *Linkages in the Landscape: The Role of Corridors and Connectivity in Wildlife Conservation (No. 1)*. International Union for conservation of nature and natural resources, Gland.

Bensaïd, S. 1995. Bilan critique du barrage vert en Algérie. *Sécheresse* 6(3), 247-55.

Besseau, P., Graham, S. and Christophersen, T. 2018. *Restoring Forests and Landscapes: the Key to a Sustainable Future. Global Partnership on Forest and Landscape Restoration*. IUFRO, Vienna.

Boedhihartono, A. K. and Sayer, J. 2012. Forest landscape restoration: restoring what and for whom? In: Stanturf, J., Lamb, D. and Madsen, P. (Eds), *Forest Landscape Restoration*. Springer, Dordrecht, pp. 309-23.

Brooks, T. M., Bakarr, M. I., Boucher, T., Da Fonseca, G. A. B., Hilton-Taylor, C., Hoekstra, J. M., Moritz, T., Olivieri, S., Parrish, J., Pressey, R. L., Rodrigues, A. S. L., Sechrest, W., Stattersfield, A., Strahm, W. and Stuart, S. N. 2004. Coverage provided by the global protected-area system: is it enough? *BioScience* 54(12), 1081-91. doi:10.1641/00 06-3568(2004)054[1081:CPBTGP]2.0.CO;2.

CBD. 1992. *UN Convention on Biological Diversity*. UN, New York.

Chazdon, R. L. and Guariguata, M. R. 2018. *Decision Support Tools for Forest Landscape Restoration: Current Status and Future Outlook* (Vol. 183). Center for International Forestry Research, Bogor.

Clewell, A., Aronson, J. and Winterhalder, K. 2004. The SER international primer on ecological restoration. *Ecological Restoration* 2, 206-7.

Clewell, A., Rieger, J. and Munro, J. 2005. *Guidelines for Developing and Managing Ecological Restoration Projects* (2nd edn.). Society for Ecological Restoration, Washington, DC.

Dudley, N., Bhagwat, S. A., Harris, J., Maginnis, S., Moreno, J. G., Mueller, G. M., Oldfield, S. and Walters, G. 2018. Measuring progress in status of land under forest landscape restoration using abiotic and biotic indicators. *Restoration Ecology* 26(1), 5-12. doi:10.1111/rec.12632.

FAO 2015. *Forest Resources Assessment*. Food and Agriculture Organization, Rome.

FAO 2016. *State of the World's Forests 2016. Forests and Agriculture: Land-Use Challenges and Opportunities*. Food and Agriculture Organization, Rome.

FAO and Global Mechanism of the UNCCD 2015. *Sustainable Financing for Forest and Landscape Restoration: Opportunities, Challenges and the Way Forward*. Discussion paper. Food and Agriculture Organization, Rome.

Gann, G. D., McDonald, T., Walder, B., Aronson, J., Nelson, C. R., Jonson, J., Hallett, J. G., Eisenberg, C., Guariguata, M. R., Liu, J., Hua, F., Echeverría, C., Gonzales, E., Shaw, N., Decleer, K. and Dixon, K. W. 2019. International principles and standards for the practice of ecological restoration. *Restoration Ecology* 27(S1), S1-S46. doi:10.1111/rec.13035.

Hallett, L. M., Diver, S., Eitzel, M. V., Olson, J. J., Ramage, B. S., Sardinas, H., Statman-Weil, Z. and Suding, K. N. 2013. Do we practice what we preach? Goal setting for ecological restoration. *Restoration Ecology* 21(3), 312-19. doi:10.1111/rec.12007.

Hanson, C., Buckingham, K., DeWitt, S. and Laestadius, L. 2015. *The Restoration Diagnostic*. World Resources Institute (WRI), Washington, DC.

Holl, K. D., Crone, E. E. and Schultz, C. B. 2003. Landscape restoration: moving from generalities to methodologies. *BioScience* 53(5), 491-502. doi:10.1641/0006-35 68(2003)053[0491:LRMFGT]2.0.CO;2.

IPBES 2018. *Summary for Policymakers of the Thematic Assessment Report on Land Degradation and Restoration of the Intergovernmental Science-Policy Platform on Biodiversity and Ecosystem Services*, Eds. Scholes, R., Montanarella, L., Brainich, A., Barger, N., ten Brink, B., Cantele, M., Erasmus, B., Fisher, J., Gardner, T., Holland, T. G., F. Kohler, J. S. Kotiaho, G. Von Maltitz, G. Nangendo, R. Pandit, J. Parrotta, M. D. Potts, S. Prince, M. Sankaran and L. Willemen. IPBES Secretariat, Bonn.

ITTO 2002. *ITTO Guidelines for the Restoration, Management and Rehabilitation of Degraded and Secondary Tropical Forests*. ITTO, Yokohama.

IUCN and WRI 2014. *A Guide to the Restoration Opportunities Assessment Methodology (ROAM)*. International Union for conservation of nature and natural resources and Washington Research Institute, Gland and Washington, DC.

Lamb, D. and Gilmour, D. 2003. *Rehabilitation and Restoration of Degraded Forests*. International Union for conservation of nature and natural resources and World Wildlife Fund, Gland.

Larson, A. J., Belote, R. T., Williamson, M. A. and Aplet, G. H. 2013. Making monitoring count: project design for active adaptive management. *Journal of Forestry* 111(5), 348-56. doi:10.5849/jof.13-021.

Löfqvist, S. and Ghazoul, J. 2019. Private funding is essential to leverage forest and landscape restoration at global scales. *Nature Ecology and Evolution* 3(12), 1612-5. doi:10.1038/s41559-019-1031-y.

Lund, J. F., Sungusia, E., Mabele, M. B. and Scheba, A. 2017. Promising change, delivering continuity: REDD+ as conservation fad. *World Development* 89, 124-39. doi:10.1016/j.worlddev.2016.08.005.

Mansourian, S. 2016. Understanding the relationship between governance and forest landscape restoration. *Conservation and Society* 14(3), 267-78. doi:10.4103/0972-4923.186830.

Mansourian, S., Razafimahatratra, A., Ranjatson, P. and Rambeloarisoa, G. 2016. Novel governance for forest landscape restoration in Fandriana-Marolambo, Madagascar. *World Development Perspectives* 3, 28-31. doi:10.1016/j.wdp.2016.11.009.

Mansourian, S., Dudley, N. and Vallauri, D. 2017a. Forest Landscape Restoration: progress in the last decade and remaining challenges. *Ecological Restoration* 35(4), 281-8. doi:10.3368/er.35.4.281.

Mansourian, S., Stanturf, J. A., Derkyi, M. A. A. and Engel, V. L. 2017b. Forest Landscape Restoration: increasing the positive impacts of forest restoration or simply the area under tree cover? *Restoration Ecology* 25(2), 178-83. doi:10.1111/rec.12489.

Mansourian, S. 2018. In the eye of the beholder: reconciling Interpretations of Forest Landscape Restoration. *Land Degradation and Development* 29(9), 2888-98. doi:10.1002/ldr.3014.

Mansourian, S., Razafimahatratra, A. and Vallauri, D. 2018. *Lessons Learnt from 13 Years of Restoration in a Moist Tropical Forest: The Fandriana-Marolambo Landscape in Madagascar*. World Wildlife Fund France, Paris.

Mansourian, S. and Parrotta, J. 2019. From addressing symptoms to tackling the illness: reversing forest loss and degradation. *Environmental Science and Policy* 101, 262-5. doi:10.1016/j.envsci.2019.08.007.

Mansourian, S. and Sgard, A. 2019. Diverse interpretations of governance and their relevance to forest landscape restoration. *Land Use Policy*, 101, 104011. doi:10.1016/j.landusepol.2019.05.030.

Margules, C. R. and Pressey, R. L. 2000. Systematic conservation planning. *Nature* 405(6783), 243-53. doi:10.1038/35012251.

McShane, T. O. and Wells, M. P. 2004. *Getting Biodiversity Projects to Work: Towards More Effective Conservation and Development*. Columbia University Press, New York.

MEA 2005. *Ecosystems and Human Well-Being: Synthesis*. Island Press, Washington, DC.

Myers, N., Mittermeier, R. A., Mittermeier, C. G., Da Fonseca, G. A. and Kent, J. 2000. Biodiversity hotspots for conservation priorities. *Nature* 403(6772), 853-8. doi:10.1038/35002501.

NEPAD 2017. *African Forest Landscape Restoration Initiative*. NEPAD, Midrand.

Olson, D. M. and Dinerstein, E. 1998. The Global 200: a representation approach to conserving the Earth's most biologically valuable ecoregions. *Conservation Biology* 12(3), 502-15. doi:10.1046/j.1523-1739.1998.012003502.x.

Ostrom, E. 2009. A general framework for analyzing sustainability of social-ecological systems. *Science* 325(5939), 419-22. doi:10.1126/science.1172133.

Pandit, R., Parrotta, J. A., Chaudhary, A. K., Karlen, D. L., Vieira, D. L. M., Anker, Y., Chen, R., Morris, J., Harris, J. and Ntshotsho, P. 2020. A framework to evaluate land degradation and restoration responses for improved planning and decision-making. *Ecosystems and People* 16(1), 1-18. doi:10.1080/26395916.2019.1697756.

Perring, M. P., Erickson, T. E. and Brancalion, P. H. S. 2018. Rocketing restoration: enabling the upscaling of ecological restoration in the Anthropocene. *Restoration Ecology* 26(6), 1017-23. doi:10.1111/rec.12871.

Pistorious, T. and Kiff, L. 2017. *From a Biodiversity Perspective: Risks, Trade-Offs, and International Guidance for Forest Landscape Restoration*. Unique Forestry and Land Use GmbH, Freiburg, Germany.

Redford, K. H., Coppolillo, P., Sanderson, E. W., Da Fonseca, G. A. B., Dinerstein, E., Groves, C., Mace, G., Maginnis, S., Mittermeier, R. A., Noss, R., Olson, D., Robinson, J. G., Vedder, A. and Wright, M. 2003. Mapping the conservation landscape. *Conservation Biology* 17(1), 116–31. doi:10.1046/j.1523-1739.2003.01467.x.

Reij, C. and Winterbottom, R. 2015. *Scaling Up Regreening: Six Steps to Success. A Practical Approach to Forest and Landscape Restoration*. Washington Research Institute, Washington, DC.

Roe, D., Nelson, F. and Sandbrook, C. (Eds) 2009. *Community Management of Natural Resources in Africa: Impacts, Experiences and Future Directions, Natural Resource Issues No. 18*. International Institute for Environment and Development, London.

Sabogal, C., Besacier, C. and McGuire, D. 2015. Forest and landscape restoration: concepts, approaches and challenges for implementation. *Unasylva* 66(245), 3.

Sasaki, N. and Putz, F. E. 2009. Critical need for new definitions of forest and forest degradation in global climate change agreements. *Conservation Letters* 2(5), 226–32. doi:10.1111/j.1755-263X.2009.00067.x.

Stanturf, J., Mansourian, S. and Kleine, M. (Eds) 2017. *Implementing Forest Landscape Restoration: A Practitioner's Guide*. IUFRO, Vienna.

Stanturf, J. A., Kleine, M., Mansourian, S., Parrotta, J., Madsen, P., Kant, P., Burns, J. and Bolte, A. 2019. Implementing forest landscape restoration under the Bonn Challenge: a systematic approach. *Annals of Forest Science* 76(2), 50. doi:10.1007/s13595-019-0833-z.

TEEB 2009. *The Economics of Ecosystems and Biodiversity for National and International Policy Makers – Summary: Responding to the Value of Nature*. TEEB, Geneva.

Thompson, I. D., Guariguata, M. R., Okabe, K., Bahamondez, C., Nasi, R., Heymell, V. and Sabogal, C. 2013. An operational framework for defining and monitoring forest degradation. *Ecology and Society* 18(2), 20–43. doi:10.5751/ES-05443-180220.

Vallauri, D., Aronson, J. and Dudley, N. 2005. An attempt to develop a framework for restoration planning. In: Mansourian, S., Vallauri, D. and Dudley, N. (Eds), *Forest Restoration in Landscapes: Beyond Planting Trees*. Springer, New York, pp. 65–70.

Wiersum, K. F. 1995. 200 years of sustainability in forestry: lessons from history. *Environmental Management* 19(3), 321–9. doi:10.1007/BF02471975.

WWF and IUCN 2000. Minutes of the forests reborn workshop in Segovia. Unpublished.

Yang, A., Bellwood-Howard, I. and Lippe, M. 2018. Social-ecological systems and Forest landscape restoration. In: Mansourian, S. and Parrotta, J. (Eds), *Forest Landscape Restoration*. Routledge, London, pp. 65–82.

# Chapter 5

## Achieving sustainable management of tropical forests: overview and conclusions

*Jürgen Blaser, Bern University of Applied Sciences, Switzerland; Patrick D. Hardcastle, Forestry Development Specialist, UK; and Gillian Petrokofsky, University of Oxford, UK*

1 Introduction: sustainable forest management (SFM)

2 Importance of tropical forest ecology

3 Forests, climate and climate change

4 Forest loss and degradation

5 Forest products and ecological services

6 Community-based forest management (CBFM) and SFM

7 SFM and the United Nations sustainable development goals (SDGs)

8 Wood products, plantations and SFM

9 Monitoring and measuring for SFM

10 Shifting cultivation and SFM

11 Forest landscape restoration (FLR)

12 SFM in major tropical regions

13 Conclusions

14 References

*'It is a truth universally acknowledged that a single country in possession of good forests, must be in want of their sustainable management.'*
*(quote adapted from Pride and Prejudice by Jane Austen)*

## 1 Introduction: sustainable forest management (SFM)

With very few exceptions, sustainable forest management (SFM) is almost universally acknowledged as being 'a good thing' for humanity and, indeed, the whole planet. Beyond this acknowledgement, however, universal agreement quickly dissolves in terms of what it means, whether it is achievable and if so, how to go about it. In this edited volume of chapters, written by subject

http://dx.doi.org/10.19103/AS.2020.0074.45

specialists from around the world, we have endeavoured to bring together a synopsis of current knowledge and thinking about different aspects of SFM to help those who are responsible for its achievement.

Sustainable timber yield has been part of forest management for more than 300 years,[1] but today, SFM casts a much wider net, including social, economic and environmental parameters, as well as inter-generational equity, in its aims of ensuring ecosystem vitality and renewal in perpetuity. Putz and Thompson have presented in their chapter a detailed discussion of SFM and highlights the challenges to its achievement. Their discussion uses hexagonal web diagrams as conceptual frameworks to exemplify the differences between five types of forest use: protection; natural forest management; plantations; community forestry; and restoration. Each hexagon covers wood products, biodiversity, carbon, water, soil and non-timber forest products (NTFPs) while the concept of sustained yield is expanded to include, for example, species composition, size distribution, quality and profitability, noting that other polygons of criteria could be created as required to clarify whether SFM is being achieved.

The core argument is that SFM is achieved through complementary aggregation of the different 'main use' hexagons within an overall landscape, and that scale is very important to the concept of SFM. Rare species require large landscapes at a scale above the minimum for survival of a specific population for sustainability, but the wider aim of SFM is to ensure each land use or forest block is sustainably managed within a wider sustainable landscape. Even exotic forest plantations can be sustainably managed with adequate attention to appropriate scale and good design. The 'Green Deserts' created by extensive blanket planting of monocultures, or extensive poorly controlled management of natural forests, create little opportunity for SFM. Opportunities for SFM are increased by carefully designed, mixed landscapes with good connectivity for biodiversity. Integrating more actively managed and protected landscapes together with adequate and effective control and good management can and does provide a basis for SFM.

## 2 Importance of tropical forest ecology

Tropical forests encompass a wide range of different forest types; consequently, the SFM of tropical forests requires knowledge and appreciation of the complexity and regeneration ecology of each type. In essence, the more complex the ecological structure and regeneration system, the more vulnerable the type

---

1 Three hundred years ago, Hans Carl von Carlowitz, a German mining administrator, was vexed by the dwindling supply of wood for the silver mines he oversaw, and he was critical about the overharvesting of the forest. He published a book, *Sylvicultura oeconomica*, in which he defined the German term for sustainability, Nachhaltigkeit and argued that the principle should be applied to the management of forests to ensure the perpetual supply of timber. He urged the adoption of measures that would make forests into a permanent economic resource (Schmithüsen, 2013).

is to both disturbance and environmental changes, as well as the more difficult its sustainable management. Forest genetics has historically been focused on planted forests, where it has resulted in major improvements to growth rates as well as enhancement of specific attributes such as wood structure and resistance to pests and diseases. New methods being developed to support forest genetics show potential application in better understanding of forest regeneration ecology as well as guiding tree improvement. The application of modern genetic methodologies and techniques to enhance understanding of complex forest ecological systems is likely to become an increasingly valuable tool for their conservation and management.

While diseases, pests and their management were generally only a topic of primary importance in plantations, these are increasingly also important issues for natural tropical forests as they become more and more affected by climate change. The effects of this are resulting in changes to the microbiological elements of forest ecosystems as well as the trees and there is useful guidance available as a result. Without application of sound knowledge of forest ecology, sustainable management will be impossible. A paper by Aubréville, in 1947, despite being written in language only acceptable at that time, argues that the most important role of dry African forests is preserving soils rather than producing timber, draws attention to the major influence of fire in forest ecology and composition, and also emphasises the importance of coordinated, international action to respond to these points.

## 3 Forests, climate and climate change

All forests grow under a continuous state of flux in response to changes in the climate and the natural cycle of maturity, senescence and rejuvenation, and these natural cycles vary in period and intensity as a result of differing levels of disturbance. Low levels of natural disturbance result in complex, relatively stable forest ecosystems that are generally the most vulnerable to change from external influences as well as being the most difficult to manage sustainably. Other highly vulnerable forests are the ones that are on the brink of survival, such as those in highly diminished niches, for example, the coniferous forest remnants on mountain refuges in dry Africa.

Different types of tropical forest have evolved in response to the characteristics of their environment and the natural agents of influence and disturbance that exist in this environment. The rapid increase in global temperature since the mid-twentieth century, as a result of the increased concentration of atmospheric greenhouse gases being much more rapid than the natural rate of change, is putting abnormal stress on all forest types and especially those most vulnerable to such change, for example, montane forests. Their natural response would be to slowly move to higher elevations but in many cases, this is physically impossible, and these forests are in substantial danger of extinction if global temperature rise is not abated.

Concern with climate change has been on the global environmental agenda since the 1980s, the three Rio conventions on biodiversity, desertification and climate change codified this concern while the importance of deforestation as a major source of greenhouse gas emissions was galvanised by the Stern report (Stern, 2014) and more recently by the IPCC report on climate change and land (IPCC, in press). The political process in the UNFCCC since the Conference of the Parties in Montreal in 2005 brought to the forefront of international dialogue on forests and climate change what is known today under the wider concept of REDD+ and stimulated attention on the need for international action to reduce forest loss and enhance carbon sinks.

Tropical forests have a vital role in the hydrological cycle as well as in the carbon cycle due to their importance as a source of atmospheric water vapour that provides rainfall; coastal deforestation in particular can be very destructive and lead to a severe climate switch from moist to dry. Given the fragility of complex forest ecosystems, especially tropical moist forests, loss or damage to these forests could lead to catastrophic loss of function in terms of both carbon and water cycles. Because tropical moist forests are generally not seriously affected by fires, excessive canopy opening that results in more flammable regrowth combined with openings for access tracks and roads can allow fire to penetrate deeply into closed forests that are highly fire-sensitive. Great care is required in management and protection to avoid these forests reaching the tipping point beyond which their role in both the carbon and hydrological cycles could be seriously undermined.

While current interest in carbon sequestration has led to improved understanding and recognition of the need for efficient supply chain use and careful management to maintain active functions, there is still quite limited understanding of the forest-climate interaction and interdependency. Further exploration and knowledge of the biotic pump theory[2] is required. If forests are to survive and continue delivering climate-related functions effectively in support of both the carbon and the hydrological cycles, there is need to invest in building adaptation and resilience to climate change into these forest ecosystems. The focus on climate change mitigation, for example, through REDD+, should also include targeted investment on building this adaptation and resilience.

---

2 The Biotic Pump theory states that whenever water vapour condenses in the atmosphere as the result of the rising and cooling down of air, energy is released that heats up the air again, reduces its pressure and accelerates air flow. Within this condensation-driven theory, the role of forests in the atmospheric circulation is active rather than passive. Increased evapotranspiration by forests leads to increased condensation, which accelerates air flows and draws in moist air from the ocean which finally leads to rainfall that surpasses local evapotranspiration.

# 4    Forest loss and degradation

In recent decades, deforestation has been primarily a phenomenon of the tropics and subtropics. While there continues to be some forest loss in temperate and boreal regions, overall, these forest areas are now stable or increasing. Their main cause of loss is wildfires in boreal forests, which accounted for one-quarter of global forest loss between 2001 and 2015, although in time much of this area may naturally regenerate. While logging is one cause, particularly of degradation rather than immediate deforestation, the main driver of tropical forest loss is conversion to agriculture. The type of agriculture, and hence reversibility of the land-use change, varies between regions.

The second quarter of global forest loss between 2001 and 2015 is ascribed to global commodity chains as a result of mining, conversion to pasture for stock rearing and clearance for commercial agricultural crops such as palm oil and soya. This land is effectively permanently lost to forest cover and restoration of mined areas is seldom undertaken in tropical regions. The loss of forest to commercial-scale farming is currently mainly in tropical America and SE Asia. A further quarter of the global forest area was lost due to subsistence farming. This may ultimately be recoverable, at least partially, depending on the type of forest converted. The final quarter of forest loss was within managed forests and plantations. This may be redeemable depending on the type of forest and whether it is then restored and protected or subjected to damaging agencies such as fire, or converted to agriculture.

Clear understanding of the causes of forest loss is an essential precursor to managing them. While conversion is often rapid, degradation and increased vulnerability to damaging agencies is usually more gradual. Disaggregated information is required to understand forest losses within individual countries and regional figures covering multiple countries can be severely misleading as natural in-growth and afforestation in one may mask serious losses in others. One critical determinant cause of forest loss is the perceived relatively low value of forest compared with that of the land on which it grows. This perception is at the root of the losses due to conversion to other uses and further exacerbated by the long time-scale necessary for most forest management processes.

# 5    Forest products and ecological services

It is useful to reflect that inadequate appreciation of currently unpriced products and service values poses a severe threat to widespread SFM because the true value of the forest, as opposed to the land on which it grows, is inadequately understood. While forest recreation and amenity are often recognised as important service values, especially when they are monetised for ecotourism, their scope is much wider; it includes unpriced cultural values and general relaxation for forest-adjacent communities, which are seldom valued or monetised. A critical point here is the lack of good data on forest service values more widely. Those estimates that have been made usually have very wide margins of accuracy. The absence of reliable information has considerable influence on higher decision-making affecting forests and often results in more clearly explicit primary aims for national economic development, such as timber production and poverty reduction, swamping recognition of other important services.

Biodiversity is clearly an important forest service and, while crucial for sound ecosystem health, its true economic value in underpinning ecological integrity still remains poorly recognised. This lack of adequate recognition remains a major threat for widespread adoption of SFM. Considering biodiversity and other forest services at a landscape level, including unpriced and directly consumed products such as NTFPs, recognising their interactions and interdependency should result in improved decision-making that is conducive to wider adoption of SFM but seldom does so. Focusing on only one or a few elements in planning and management, without consideration of the interactions across the whole range of services and functions can lead to perverse and damaging outcomes. This is particularly relevant to those forests managed with an excessively narrow focus on only, for example, timber extraction or, more recently, only carbon.

Maintaining sound biodiversity is crucial to retaining fully all forest functions and consequently their sustainability. There are important linkages between forest biodiversity and that on surrounding land under other uses, which again points up the benefits of a landscape-level perspective in working towards SFM. While fragmented forests may still maintain some biodiversity functions, depending on the size and spatial proximity of the fragments, highly fragmented forests are in danger of becoming unsustainable islands in the long term as many tree species may fail to pollinate adequately when biodiversity is heavily degraded. The full impact of biodiversity losses on ecological function is still poorly known and understood but in due course might rank among the major global drivers of environmental change.

While wood may be the major economic and traded product, NTFPs are also important. NTFPs are often directly consumed rather than traded, provide a valuable safety net for forest-dependent communities and, at times, others. In the late 1980s, there was great interest in commercialising NTFPs as a way of increasing the value of the forest to try and secure its retention. Ultimately, this proved generally unsatisfactory, not least because many NTFPs have limited shelf life and often ephemeral market opportunities. NTFPs are, however, important props for forest conservation and, as commercialisation is likely to lead to overharvesting, this could destroy unwittingly the fragile web of life in complex forest ecosystems.

Forest-dwelling and forest-dependent local communities generally value their forests and it is often external users and/or influences that damage the forest rather than these communities themselves, especially where they are long-established. Strict protected areas that do not allow extractive use restrict access even for limited NTFP collection. Ultimately, the best sources of sustainable supplies of NTFPs often come from mixed, locally owned agroforest plots and homegardens. Complex homegardens are most successful in wetter climates and are generally much less productive in dry ones. The latest knowledge of ancient human use in closed and apparently largely undisturbed forests suggests that even closed tropical forests (Amazon, Congo Basin) with few, or in places no, humans living there currently show clear evidence of past human occupation and use. At present the role of human activity in forest ecology and succession is unclear.

## 6 Community-based forest management (CBFM) and SFM

Although basic forest products such as fuelwood, food and medicinal plants still feature strongly in most rural African livelihood strategies, pressures for conversion to agriculture continue, driven by poverty and low agricultural productivity. Poorly defined access and tenure rights are often exacerbated by the issue of poor people living adjacent to productive forests in which they have no stake, and from which they derive little benefit beyond occasional employment. One method of bridging this gulf is community-based forest management (CBFM)[3] in which local community members are actively engaged and their needs and wishes made explicit. CBFM is complex and requires strong institutional support as well as inputs such as training, advice and oversight. This leads inevitably to the need for consistent support over considerable periods to be fully effective, as there are nearly always complex interactions around forests covering varying interests in the forest itself as opposed to the underlying land, with resultant pressures for conversion as well as access.

Community engagement through CBFM can be a valuable strategy for national SFM. Five key factors have been identified affecting the success of community forestry: the importance of effective governance; secure property rights; social equity; government support; and tangible benefits. To these can be added four key and salutary points: power is seldom shared voluntarily; CBFM typically requires long-term capacity building; there is a need for those engaged to have both short-term cash income and longer-term material benefits; and land and tree tenure are probably the most complex issues determining CBFM success.

---

3 CBFM is also known as joint forest management, collaborative forest management or participatory forest management. There are differences between these, but fundamentally all apply the same concepts. CBFM is used here to encompass all these approaches.

If CBFM is to succeed, it is vital to recognise different forest aims, such as the production of timber or a more widely based range of products and services. Concepts such as criteria and indicators provide a useful guide to the sound practices required for SFM of CBFM. Of crucial importance also is sound and effective governance to control elite capture, provide clear rights and ownership and support good practices if the African forest resource is not to be largely lost. Where clear rights and tenure are absent, deforestation and degradation tend to be more prevalent.

According to the FAO (2020), only 5% of forests in East, West, Central and southern Africa are privately owned although accurate data is lacking. In South America, Amazon countries had, by 2017, transferred between one-third and over one half of forest land to indigenous peoples/local community (IP/LC) ownership or at least partially done so. In Asia, one-third of the forest area is devolved to local communities, but this is mainly in China. Some 15% of the forest area in India is privately owned, but in other countries it is almost entirely state owned. In SE Asia and Oceania, practically all forests in PNG are communally owned while 35% of those in the Philippines are community owned or designated for their use. In Indonesia by 2017, only a very small portion was designated for IP/LCs, far less than the privately owned forest area.

Progress with agreements on IP/LC control and/or ownership since 2017 has been generally slow, despite numerous national and international court judgements. There is some evidence for a hesitant impetus that may ultimately lead to wider recognition of IP/LC rights, but this remains fragile. There is a lack of good data on tenure and rights, but decisions made in international conventions and rising interest in the issue by bilateral and multilateral donor agencies may create forward movement and support the small steps being taken by a number of countries.

## 7    SFM and the United Nations sustainable development goals (SDGs)

Using the lens of the UN sustainable development goals (SDGs) as a narrative is helpful in bringing particular focus on the direct consumption of forest products and services as part of forest-dependent communities' livelihood strategies, as well as use by more distant communities as part of their responses to poverty and food insecurity. These forest products and services important for livelihoods are mainly unpriced and unvalued, reinforcing the earlier observation that top-level government planning with its focus on increased economic production and poverty reduction often overlooks the value of these.

It is hard to generalise the progress made towards SFM other than to observe that wealthier countries able to devote human and financial resources and that are less, or not at all, dependent on external financing, have generally made more and more rapid progress. The UN SDG goals provide a valuable framework for guidance on securing more widespread SFM in poorer countries through more extensive use of community engagement, ensuring secure and equitable tenure and long leases where forest management is devolved. Cross-sectoral approaches are also essential, since the drivers of forest loss lie mainly outside the forest sector itself, while a strong focus on 'preservation' rather than sustainable use is often counter-productive. Clear and sound land-use planning leading to well-managed, productive natural forests and plantations, combined with support for good agroforestry practices can together generate employment and add value, thus addressing the key needs of improved economic growth and poverty reduction.

## 8  Wood products, plantations and SFM

In parallel with greater awareness of the role of forests in providing climate-moderating services, there have been major changes in wood-using technology in the past decades. Globalisation has driven the commoditisation of many wood products, which places downward pressure on prices and increases price sensitivity. Rapidly evolving engineered-wood technologies make use of cheaper, small-sized material and increasingly a mix of wood-based feedstock combined with other biomass and inert mixers. As industrial processes mostly require consistent raw material of regular sizes, and given the effects of globalisation, tree plantation sources are increasingly preferred. As well as increasing demand for solid and engineered wood products, there is also continuing growth in demand for packaging material. The result of these changes will be to favour and encourage increased tropical plantations at a range of scales to provide the raw material.

Because plantations are able to provide high levels of production of core products compared with what is possible from most natural forests, even if intensively protected and managed, they have an important role to play in achieving SFM. While plantations at a range of scales can reduce pressure on natural forests, there must also be good plantation design and operating standards if their role in SFM is not to be prejudicial. Both of these requirements need sound policies and effective enforcement. Even where these do exist, they are not always implemented adequately in tropical countries.

In 2015, tropical and subtropical plantations totalled 278 million ha, roughly one-third of the world's total, which itself is 7% of global forest cover. Plantation wood provides 50% of industrial roundwood and both the demand for this and the proportion that is plantation grown are likely to increase. Whilst planted forests are not inherently bad, forest plantation practices have not always been good, let alone sustainable. The bulk of tropical and subtropical plantations relies on a few species from three genera - *Acacia*, *Eucalyptus* and *Pinus*. These three genera are used because they respond well to being grown in plantations and they produce raw material that meets the requirements of the processing industries they support. If plantations are to be sustainable, then as well as the rate of wood production, there must be consideration of nutrient cycling, biodiversity, ecosystem services and social impact; these are core elements in SFM. Advances in tree improvement and plantation silviculture have largely overcome issues of declining yield, and average productivity per unit area continues to increase to many times the timber harvest volume achievable from managed natural forests. Other considerations for SFM of plantations start from the previous land use prior to planting; converting good quality, intact natural forests to plantations is counter to the principles of SFM.

Biodiversity may be temporarily depleted during plantation establishment, but thereafter if it is affected by the scale of operation (smaller is preferable), the species planted (heavy shade tends to suppress ground vegetation) and the intensity of management once the plantation is established. Good design can overcome many negative biodiversity impacts but note here also the importance of connectivity rather than overall surface area. Social impacts related to the specific land area and wider land-right claims have proven particularly potentially problematic for some plantation programmes. New plantations can create good social benefits with employment, business opportunities and so on. Careful timing and sequencing are required to sustain these benefits. Many large plantation projects were established very quickly and then left with low levels of input, which prejudiced a continuous flow of benefits. Highly productive wood plantations usually use materials that require high inputs and knowledgeable management. While this is good for specialist companies and effective public institutions, block transfer of the same technology packages to smaller growers, communities or farmers, may not be suitable if they lack skills, knowledge and resources. Plantations need to be based on site-user-species matching to accommodate this.

Decision-making at all levels from the strategic to ground level in managed natural forests and plantations will nearly always require some use of economic tools; such decisions require adequate data if they are to be sound and appropriate. It has already been emphasised that failure to recognise the true value of forest products and services, whether or not these are monetarised, often leads to poor decisions. Because wood has historically been the dominant priced forest product, managed natural forests that have low wood-production potential become highly vulnerable to conversion as liquidation of the remaining forest and the near-immediate cash flow from the new land use have high apparent value.

Plantations, financed either through public or commercial funding, are nearly always subjected to investment economic analysis using discounting, the reverse of compound interest, in which distant values are reduced geometrically over time. It is inherent in the computation formula that the highest Internal Rate of Return (IRR) is achieved by a combination of low establishment cost, high productivity and short rotations. Unless care is taken, this inevitably leads to large blocks of monocultures of quickly maturing species of commercial value to aid industrial efficiency by providing uniform raw material as close as possible to the point of transformation, the 'Green Deserts' noted earlier. The same mathematical logic makes liquidating forest to more than cover the cost of establishing oil palm plantations very attractive to the unscrupulous. Carefully crafted policy and fiscal measures encompassing well-designed regulations and fiscal support combined with clearly defined operational standards plus the capacity and willingness to enforce these effectively are essential to direct decision-making and forest management practices along desirable pathways.

Concern over the loss of tropical forests due to excessive timber harvesting resulted in the concept of certification, stimulated by the increasing general concern with the environment in the 1980s, which culminated in the Earth Summit in 1992 and its three conventions on climate change, biodiversity and desertification. Initially, certification met with strong resistance to the underlying concept and it was consumer support, especially in European and North-American markets, that enabled it to advance. Certification schemes rely on principles, criteria and indicators. Over the years, the range and depth of these core tools have both expanded to cover social matters such as land- and forest-use rights for indigenous and forest-dependent communities. As well as a certification tool, this structure can be usefully applied in framing CBFM.

Like any commercial audit system, certification requires willing buyers of the service in order to continue to be viable; this can lead to pressures on what is supposed to be an independent system. The most successful schemes have been those that are international or at least regional in application. Despite having its genesis as a system for managed natural tropical forests, certified forests are predominantly in Europe and North America, and include a significant area of planted forests. The key point is that, while certification can provide recognition of good management, it provides no bulwark for those forests where the owners and/or managers have little or no interest in sustainable forest management. Certification is a valuable tool for attracting green investors and for where markets are willing to pay for ethical standards; it cannot effect change where forest owners and managers are resistant.

## 9 Monitoring and measuring for SFM

Remote sensing technology is a major tool for monitoring loss of forest cover and, increasingly, for monitoring degradation within forests. The advent of vastly increased computing power harnessed to artificial intelligence, machine learning and improved remote sensing technologies suggests that this field will become increasingly important and useful to guide decision-making on land-use options as well as on those relating directly to forest management. For smaller and fragmented forests, ground-based measurements will continue to have an important role to play. An important issue for many tropical countries is whether the introduction of ever more sophisticated technology will further widen the emerging technology gap, especially for the poorest countries that often have very limited IT capabilities and infrastructure, thereby fuelling further growth in the 'monitoring industry' that is largely the preserve of more developed economies. Currently, Africa seems to be particularly vulnerable in this respect.

National reporting on progress with climate change has become progressively more important within the international arena and for securing financial support. Advances in monitoring and reporting of emissions from tropical forests have made it increasingly complex, again reflecting the disadvantages for poorer countries with limited access to qualified personnel and advanced technology. It is also observed that there is widespread use of 10% canopy cover to determine the extent of forests. Canopy cover of 10% in what was originally closed rainforest represents pretty close to annihilation whereas in a dry savanna woodland, it would be quite acceptable or even gratefully received.

## 10    Shifting cultivation and SFM

Shifting cultivation systems can be a form of SFM. Trees are left for many purposes related to direct livelihood support and, sometimes, to aid restoration of fertility for the next cropping cycle. While return cycles were long historically, allowing the forest adequate time to recover, these cycles have generally become much shorter. In savanna ecosystems, tree regeneration by suckering may be more important than by seed, which can be helpful to rapid regrowth. The length of the return cycle affects regrowth, fertility restoration and service values including carbon and biodiversity. Increasing intensity of cultivation as well as shortening of return cycles reduces subsequent woody regrowth, carbon stocks and biodiversity.

The length of the return cycle in shifting cultivation is very important in terms of biodiversity. Even return cycles close to 50 years only restore some 80% of moist forest biodiversity. Spatial distribution is also important, while distance from the forest edge is strongly linked with invertebrate richness, meaning scattered small, individual plots have much less loss compared with large clearings. Intensified shifting agriculture through larger areas and/or reduced return cycles will ultimately be destructive of the structure, volume, biodiversity and carbon storage of the forest and also reduce the direct benefits to the practitioners. These points are of considerable relevance to CBFM, which often includes managed natural forests and woodlands as well as other systems, and also relevant to restoration.

## 11  Forest landscape restoration (FLR)

Forest landscape restoration (FLR) is a process that restores in parallel both ecological integrity, which was the original driver for its development, and human benefits. In its application, building connectivity is more important than blanket restoration of large areas and it can include a range of land uses at site level as well as involving building resilience to reduce the risk of the cycle of forest loss restarting. In general, supportive FLR policy aims are not currently being fully complemented by consideration of the trade-offs required and more attention is needed to the resolution of the potential conflicts that will inevitably occur as a result of this. Implicitly, effective FLR must have good stakeholder engagement at the initial stage and also for refining the detail of activities. The process itself has to be supported by sound policies and strategies, effective institutions and adequately enforced operating standards.

One vital point to make in respect of FLR is that, unless there is adequate attention given to resolving the drivers that initially led to forest loss and/or degradation, there is nothing to ensure that this will not simply happen again. In particular, restored forests must be proofed against capture by elites or vested interests. Effective FLR needs time as well as political support, financial and human resources. In summary:

> On the one hand, increased global interest has provided an entry point for FLR interventions and helped to raise funding for these. On the other hand, it has put pressure to achieve quick results, with very limited understanding of the complexity of the issues, the timeframes involved and the need for multiple level interventions. This gives rise to the challenge: How can the willingness to pledge large scale funding be translated into a realistic plan for effective delivery on the ground?

Time frame is one key consideration but so are actual and adequate capacity, commitment and realism in FLR countries, as well as among donors.

## 12  SFM in major tropical regions

Before considering progress with SFM in various tropical regions, it is interesting to look at a basic comparison by region of the sustainable timber yield currently accruing in the forest resource compared with the volume harvested from that forest resource, excluding illegal cutting and clearance. West Africa, Central America, Asia and Oceania are generally heavily overharvested. In Central Africa, the harvest volume overall exceeds the volume of sustainable yield while in South America, harvested volume and sustainable timber yield are more or less in balance.

## 12.1 The Congo Basin

Analysis of forest management in the Congo Basin concludes that, while some areas show positive results, overall, SFM is still the exception. While the severe loss of forest cover in Ghana and Côte d'Ivoire has been driven by agricultural expansion, especially of cocoa, Congo Basin countries are currently seeking industrial agriculture as a growth strategy, including in particular oil palm plantations. Underlying this picture is the projected increase in the current population, which is expected to more than double by 2050. Added to this is limited infrastructure and political instability in some countries, leading to high numbers of displaced persons across the region, who then rely on the local forest resources in their temporary refuge for fuel and building material.

There is very limited devolution of ownership, heavy reliance on concessions for forest management and the political attraction of mining and commercial agriculture to fund economic growth. Across the region, there is a huge gap between the theory of land-use zoning, forest management plans and sustainable harvesting limits on the one hand and reality on the ground on the other. Land-use zoning is still largely incomplete, certainly in terms of legality, and most Congo Basin countries have substantial overlapping zones for forestry and other alternative land uses that, if applied, would eliminate forest cover. Management plans are seen as the basis for guaranteeing SFM and these should be prepared by governments, or at least subjected to close scrutiny prior to approval. In most cases, concession holders make the plans and government agencies approve them. Only Central African Republic has indefinite concessions, in other countries the concession period is from 15 to 30 years, the uncertainty from which limits investment and motivation for sustainable practice. There are certainly some areas of good practice, mainly involving companies that have, or are seeking, certification so they can access more discriminating markets. At present, only 10% of the overall concession area of 50 million ha (20% of the overall area with management plans) is certified.

This fragmented approach means there is lack of a clear vision of the national forest sector in each country. There is under-investment in processing and worked-out concessions are reallocated with new management plans that try to exploit what remains. In general, the result is a lack of continuity of management objectives and hence no consistent practice being applied. Where short-term harvesting concessions are offered in parallel with full concessions that at least nominally require SFM, and often on similar terms, this disparity undermines motivation for good practice unless market requirements so dictate. Despite the substantial investment in policy development, through regional and national initiatives, there is evidence of policy failure in reality. Linked to this is the low level of trained and competent personnel in the field, with individuals responsible for 30 000-50 000 ha of concessions and twice that area if all forest is included. In many cases, personnel also lack the financial and physical resources required for them to function efficiently and effectively.

There is much talk and written material about SFM, including in policies and laws, but application is extremely limited. New technologies are available and offered, but in practice the ability of most of the personnel to make use of these is equally limited. The overall picture in the Congo Basin suggests that, unless major changes at wholesale scale are made quickly, the area is likely to go the same way as West Africa, leading to large areas of forest being lost through conversion to other land uses.

## 12.2 The Brazilian Amazon

The Brazilian Amazon contains the world's largest block of rainforest, which influences global climate regulation as well being of national environmental importance and a source of national wealth. The country has a strong institutional framework for supervision of the complex laws and regulations that aim to control forest harvesting and implement SFM through detailed management plan and harvesting limits. SFM is based on polycyclic felling and, as in Congo Basin, there are concerns that the return cycle may be inadequate for some species. Relatively small trees, low frequency of desirable species and limited extraction volumes make covering the costs while complying with all harvesting regulations challenging. Even with Reduced Impact Logging (RIL), damage rates remain high and future supplies of some of the most desirable species are uncertain.

In recent decades, there has been enhanced control of illegal logging and illegal mining and recognition of community rights enabling community management over part of the forest area. Despite this being legally possible, putting community management into practice has proven complex and slow. Substantial natural forest areas are also privately owned and their management still has echoes of past times where forest exploitation was an essential precursor to securing ownership. Public land is owned by both state and federal governments. There is devolution of control over state lands to the state level. Management and harvesting operations are mainly conducted by concession holders but the concession confers no additional rights, such as payments for carbon sequestration, other than to harvest timber. Overall, while there is good potential for SFM in Brazil, especially through CBFM, numerous constraints prevent its rapid and effective implementation. These include the complexity of laws, regulations and agency mandates, lack of knowledge on the part of communities and also the continual pressure for conversion to other land uses to secure high initial returns from this.

## 12.3 Dry African forests

While both the Congo Basin and the Amazon are predominantly home to tropical moist forests, tropical dry forests occupy a substantial area in Africa and are very important for the livelihoods of a substantial proportion of the population. In Africa's dry forests, the widespread system of controlled early burning that occurred in most of these forests until the 1970s has largely fallen into disuse with predictable effects. The main cause of loss of dry forests in Africa has been due to agricultural expansion driven by population growth with little intensification of agriculture to accommodate this increase. Many of the products and services from dry woodlands are directly consumed and most are not monetarised, leading to their often being overlooked in higher-level planning and decision-making. Remedying this oversight is a key priority. However, it is also important to draw attention to the urgent need for agricultural intensification to take the pressure off continued dry forest conversion and re-institute the key basic management tools of controlling fire and grazing, preferably through CBFM.

## 13   Conclusions

Summarising the regional analyses and expanding them to all tropical forest biomes, the following overall assessment and conclusions can be drawn in respect of the future for sustainable management of tropical forests:

- In the tropical moist biome, population and income growth will heavily influence land and forest use, particularly in Africa and SE Asia. It can be expected that considerable parts of the tropical moist forests in the Congo Basin, which are relatively accessible, will be converted to agricultural land. The Amazon Basin, the Mekong and some of the major islands of Indonesia will also experience considerable forest loss in the coming 50 years or so to make way for commercial crops to meet worldwide demand for food, fodder and bioenergy;
- Climate change will become a major issue in these regions, not only for forests but also for agricultural production. Biodiversity and habitat loss will accelerate and there is a risk of complete land degradation, particularly in the Congo Basin, where a savanna/forest mosaic could become the dominant landscape feature, and in lowland SE Asia;
- In the longer term, it is expected that large-scale reforestation will take place predominantly in the tropics, rather than in temperate and boreal

regions, where fast-growing tree species can sequester carbon and produce fibre more rapidly; and

- Tropical dry biomes are likely to have different pathways. Some regions will receive more precipitation and humidity (e.g. the Sahel), and some will be more at risk of extended drought due to changing atmospheric circulation (e.g. the monsoon areas of eastern Africa and India). Semi-arid and semi-humid tropical forests, including those on the Indian subcontinent and in parts of Central America and southern South America, and mountain forest ecosystems will be among the most vulnerable forest ecosystems, due to extreme events. Overall, tropical dry biomes will expand in area, but tree cover is likely to reduce.

Finally, considering the various insights and perspectives that emerge from the chapters in this book, a number of key points can be drawn:

- Despite slow progress, setbacks and disappointments, there is strong body of opinion among forestry professionals and those in complementary disciplines that SFM is a valuable goal and that it is achievable;
- SFM has clearly entered the agenda of many international conventions and seeking its achievement has attracted substantial amounts of funding from a range of initiatives to which forests are relevant, although the time scale within which solid results can be achieved is often considerably longer than the financing cycle;
- Inadequate recognition of the true economic value of directly consumed forest products and of unpriced, and indeed often unrecognised, forest ecosystem services, leads to decision-making based on incomplete understanding of the long-term value of tropical forests. The consequence of this is over-exploitation and conversion to other uses;
- Ensuring the stability of forest carbon stocks is a major challenge for SFM. There are high expectations that REDD+ will become a major tool for funding forest management. However, there is considerable work to do to put this, or other similar mechanisms, such as those promoting payments for ecosystem services, into effect;
- As primary forests need to be carefully managed mainly for protective purposes, managing degraded and secondary forests will become more important. In combination with enrichment planting, this might lead to new forms of short-rotation forestry in the tropics, where a balance of biodiversity conservation and biomass productivity/production can be achieved;
- There are many legitimate concerns about the potential harmful ecological and social impacts of forest plantations, including at times resistance to the use of exotic species, but sufficient experience has been now accumulated

to be able to avoid such negative impacts in the future. As wood is the only truly sustainably produced material for many purposes, forest plantations for the production of wood will be needed to meet the expected increases in demand, and this increase brings with it the co-benefits of non-wood goods and services, including watershed and soil protection, recreation and carbon sequestration;

- Much decision-making related to forests is based on inadequate and incomplete data and the lack of use of existing information. Science/policy interfaces are well established for broader environmental concerns, such as climate change and biodiversity, but are today almost completely lacking with respect to SFM in the tropics;
- While specific interests such as: REDD+ to mitigate climate change; a focus on forest law enforcement, governance and trade to address illegal logging; and forest landscape restoration have all captured political interest and hence finance, much of the world's tropical forests still remain inadequately protected and managed due to the failure to bring to bear an adequate level of resources to ensure their day-to-day protection and careful management;
- Although forests are intimately and inextricably entwined in the livelihoods of a substantial proportion of the human population in tropical countries, their voices are often given inadequate weight. As a result, external influences, driven by short-term profit or short-term benefits that accrue only to the external parties have been allowed to flourish;
- Real and effective CBFM, supported by a knowledgeable and creative cadre of well-resourced personnel must have a much more prominent role in seeking SFM of tropical forests than it has been allowed to have so far; and
- As natural forests in the tropics become more vulnerable and fragile due to the fast pace of change, especially climate change, maintaining the production of forest goods and ecosystem services will likely depend increasingly on human interventions and ingenuity. In this regard science and governance reform will both have important roles to play to achieve SFM.

With their huge protective and productive functions, forests, and in particular tropical forests, must continue to play a crucial global role today and in the future. Knowledge of the art and practice of sustainably managing tropical forests will be in high demand by society. As one of the main renewable natural resources, forests will be expected to help mitigate climate change, protect soil and water, provide clean air, conserve biodiversity, help maintain the mental and emotional health of humans who live in them and who visit them to find space and calmness. Concurrently, these forests will also be required to produce wood fibre and other products. This is a tall order but with adequate recognition, finance and human resources made available, it can be achieved through adoption of the principles and practice of SFM.

Humanity's future will depend in large measure on how successfully it deals with its forests by stopping forest loss and increasing tree-covered areas. There is still time to do this and the ability is there to implement SFM, but while today's and tomorrow's foresters have much work to do, they can only achieve success with wider political, economic and social support. The precautionary principle must be more strongly used in decision-making affecting forests to support foresters and others seeking to achieve SFM.

What Aubréville observed as long ago as 1947 still remains true today: 'Forest masses yield an influence over wide distances and what is done in one country may have repercussions in other distant countries. Pooling of effort, joint research, and the synchronizing of undertakings appear to us imperative if failure is to be avoided.'

# 14  References

Aubréville, A. M. A. (1947). The Disappearance of the Tropical Forests of Africa. *Unasylva* 1(1):5–11. Reprinted with minor revisions in *Fire Ecology* Volume 9, Issue 2, 2013. doi: 10.1007/BF03400624.

FAO (2020). *Global Forest Resources Assessment 2020*. FAO, Rome. doi: 10.4060/ca9825en.

IPCC (in press). Summary for Policymakers. In: *Climate Change and Land: an IPCC Special Report on Climate Change, Desertification, Land Degradation, Sustainable Land Management, Food Security, and Greenhouse Gas Fluxes in Terrestrial Ecosystems* [P. R. Shukla, J. Skea, E. Calvo Buendia, V. Masson-Delmotte, H.-O. Pörtner, D. C. Roberts, P. Zhai, R. Slade, S. Connors, R. van Diemen, M. Ferrat, E. Haughey, S. Luz, S. Neogi, M. Pathak, J. Petzold, J. Portugal Pereira, P. Vyas, E. Huntley, K. Kissick, M. Belkacemi, J. Malley, (eds.)]. Available at: https://www.ipcc.ch/site/assets/uploads/2019/11/02_Summary-for-Policymakers_SPM.pdf.

Schmithüsen, F. (2013). Three Hundred Years of Applied Sustainability in Forestry. *Unasylva* 64(240):3–1.

Stern, N. 2014. *The Economics of Climate Change*. Cambridge University Press, Cambridge. doi: 10.1017/CBO9780511817434.